T0332547

VLSI FOR ARTIFICIAL INTELLIGENCE

THE KLUWER INTERNATIONAL SERIES
IN ENGINEERING AND COMPUTER SCIENCE

VLSI, COMPUTER ARCHITECTURE AND
DIGITAL SIGNAL PROCESSING

Consulting Editor

Jonathan Allen

Other books in the series:

Logic Minimization Algorithms for VLSI Synthesis. R.K. Brayton, G.D. Hachtel,
C.T. McMullen, and A.L. Sangiovanni-Vincentelli. ISBN 0–89838–164–9.
Adaptive Filters: Structures, Algorithms, and Applications. M.L. Honig and
D.G. Messerschmitt. ISBN 0–89838–163–0.
Introduction to VLSI Silicon Devices: Physics, Technology and Characterization.
B. El-Kareh and R.J. Bombard. ISBN 0–89838–210–6.
Latchup in CMOS Technology: The Problem and Its Cure. R.R. Troutman.
ISBN 0–89838–215–7.
Digital CMOS Circuit Design. M. Annaratone. ISBN 0–89838–224–6.
The Bounding Approach to VLSI Circuit Simulation. C.A. Zukowski.
ISBN 0–89838–176–2.
Multi-Level Simulation for VLSI Design. D.D. Hill and D.R. Coelho.
ISBN 0–89838–184–3.
Relaxation Techniques for the Simulation of VLSI Circuits. J. White and
A. Sangiovanni-Vincentelli. ISBN 0–89838–1ö5–X.
VLSI CAD Tools and Applications. W. Fichtner and M. Morf, editors.
ISBN 0–89838–193–2.
A VLSI Architecture for Concurrent Data Structures. W.J. Dally. ISBN 0–89838–235–1.
Yield Simulation for Integrated Circuits. D.M.H. Walker. ISBN 0–89838–244–0.
VLSI Specification, Verification and Synthesis. G. Birtwistle and P.A. Subrahmanyam.
ISBN 0–89838–246–7.
Fundamentals of Computer-Aided Circuit Simulation. W.J. McCalla. ISBN 0–89838–248–3.
Serial Data Computation. S.G. Smith and P.B. Denyer. ISBN 0–89838–253–X.
Phonological Parsing in Speech Recognition. K.W. Church. ISBN 0–89838–250–5.
Simulated Annealing for VLSI Design. D.F. Wong, H.W. Leong, and C.L. Liu.
ISBN 0–89838–256–4.
Polycrystalline Silicon for Integrated Circuit Applications. T. Kamins.
ISBN 0–89838–259–9.
FET Modeling for Circuit Simulation. D. Divekar. ISBN 0–89838–264–5.
VLSI Placement and Global Routing Using Simulated Annealing. C. Sechen.
ISBN 0–89838–281–5.
Adaptive Filters and Equalisers. B. Mulgrew, C.F.N. Cowan. ISBN 0–89838–285–8.
Computer-Aided Design and VLSI Device Development, Second Edition. K.M. Cham,
S-Y. Oh, J.L. Moll, K. Lee, P. Vande Voorde, D. Chin. ISBN: 0–89838–277–7.
Automatic Speech Recognition. K-F. Lee. ISBN 0–89838–296–3.
Speech Time-Frequency Representations. M.D. Riley. ISBN 0–89838–298–X
A Systolic Array Optimizing Compiler. M.S. Lam. ISBN: 0–89838–300–5.
Algorithms and Techniques for VLSI Layout Synthesis. D. Hill, D. Shugard, J. Fishburn,
K. Keutzer. ISBN: 0–89838–301–3.
Switch-Level Timing Simulation of MOS VLSI Circuits. V.B. Rao, D.V. Overhauser,
T.N. Trick, I.N. Hajj. ISBN 0–89838–302–1

VLSI FOR
ARTIFICIAL INTELLIGENCE

edited by

José G. Delgado—Frias

Department of Electrical Engineering
State University of New York at Binghamton

Will R. Moore

Department of Engineering Science
University of Oxford

KLUWER ACADEMIC PUBLISHERS
BOSTON/DORDRECHT/LONDON

Distributors for North America:
Kluwer Academic Publishers
101 Philip Drive
Assinippi Park
Norwell, Massachusetts 02061 USA

Distributors for the UK and Ireland:
Kluwer Academic Publishers
Falcon House, Queen Square
Lancaster LA1 1RN, UNITED KINGDOM

Distributors for all other countries:
Kluwer Academic Publishers Group
Distribution Centre
Post Office Box 322
3300 AH Dordrecht, THE NETHERLANDS

Library of Congress Cataloging-in-Publication Data

VLSI for artificial intelligence.

(The Kluwer international series in engineering and
computer science ; 68)
    Includes bibliographies and index.
    1. Artificial intelligence—Data processing.
2. Integrated circuits—Very large scale integration.
I. Delgado-Frias, José G.   II. Moore, Will R.
III. Series.
Q336.V57   1989   006.3   88-37254
ISBN 0-7923-9000-8

Printed in the United States of America

Contents

List of Contributors

T Ae, *Hiroshima* (Japan)
R Aibara, *Hiroshima* (Japan)
L A Akers, *Arizona State* (USA)
I Aleksander, *Imperial College* (UK)
S Bandyopadhyay, *Windsor* (Canada)
Y Bekkers, *INRIA* (France)
F Blayo, *LGI* (France)
M Brule, *Syracuse* (USA)
Z F Butler, *Edinburgh* (UK)
L Chevallier, *INRIA* (France)
C L Chng, *Nanyang* (Taiwan)
P Civera, *Torino* (Italy)
J G Delgado-Frias, *Oxford* (UK)
S Delgado-Rannauro, *Essex* (UK)
D Del Corso, *Torino* (Italy)
R Eck, *Erlangen-Nuernberg* (FRG)
B Faure, *IMAG* (France)
D K Ferry, *Arizona State* (USA)
R A Frost, *Windsor* (Canada)
S Garth, *Texas Instruments* (UK)
R Ginosar, *Technion* (Israel)
R J Glover, *Brunel* (UK)
K Goser, *Dortmund* (FRG)
R O Grondin, *Arizona State* (USA)
A Harsat, *Technion* (Israel)
P Hurat, *LGI* (France)
P G A Jespers, *Louvain* (Belgium)
S Kak, *Louisiana State* (USA)

P M Kogge, *IBM* (USA)
S D Krueger, *TI* (USA)
S H Lavington, *Essex* (UK)
S Le Huitouze, *INRIA* (France)
G Mazare, *IMAG* (France)
K-Y Mok, *Essex* (UK)
W R Moore, *Oxford* (UK)
A F Murray, *Edinburgh* (UK)
Y H Ng, *Imperial College* (UK)
J T O'Donnell, *Glasgow* (UK)
J Oldfield, *Syracuse* (USA)
D Phoukas, *Windsor* (Canada)
D Pike, *Cambridge* (UK)
G L Piccinini, *Torino* (Italy)
D Prados, *Louisiana State* (USA)
J Reynolds, *Essex* (UK)
O Ridoux, *INRIA* (France)
I N Robinson, *HP* (USA)
J Robinson, *Essex* (UK)
U Rückert, *Dortmund* (FRG)
B Sirletti, *Louvain* (Belgium)
A V W Smith, *Edinburgh* (UK)
C Stormon, *Syracuse* (USA)
L Tarassenko, *Oxford* (UK)
L Ungaro, *INRIA* (France)
M Verleysen, *Louvain* (Belgium)
M R Walker, *Arizona State* (USA)
M Weinfeld, *Polytechnique* (France)
M Zamboni, *Torino* (Italy)

Preface

This book is an edited selection of the papers presented at the *International Workshop on VLSI for Artificial Intelligence* which was held at the University of Oxford in July 1988. Our thanks go to all the contributors and especially to the programme committee for all their hard work. Thanks are also due to the ACM–SIGARCH, the Alvey Directorate, the IEE and the IEEE Computer Society for publicising the event and to Oxford University for their active support. We are particularly grateful to David Cawley and Paula Appleby for coping with the administrative problems.

José Delgado–Frias
Will Moore

October 1988

Programme Committee

Prologue

Research on architectures dedicated to artificial intelligence (AI) processing has been increasing in recent years, since conventional data– or numerically–oriented architectures are not able to provide the computational power and/or functionality required. For the time being these architectures have to be implemented in VLSI technology with its inherent constraints on speed, connectivity, fabrication yield and power. This in turn impacts on the effectiveness of the computer architecture.

The aim of this book is to present the state–of–the–art and future trends on VLSI implementations of machines for AI computing. In order to achieve this objective the papers are drawn from a number of research communities spanning the subjects of VLSI design through computer architectures to AI programming and applications.

This book has eight chapters which have been grouped into three major categories: *hardware support for artificial intelligence programming languages, computer architectures for knowledge oriented systems,* and *neural network hardware implementations.* This grouping covers the complete range from purely programmable systems to learning systems and from symbolic manipulation to connectionism.

Hardware support for artificial intelligence programming languages

Logic–oriented programming languages –such as Prolog– and functional languages –such as pure Lisp and Miranda– have been widely used as high–level languages for artificial intelligence applications. As a consequence, much research has been carried out to develop high performance computers for these programming languages. **Chapter 1** contains papers which examine the implementations of Prolog machines. Although, the majority of these machines are based on the Warren abstract machine (WAM), there is a wide range of architectures: from reduced instruction set computers (RISC) to complex instruction set computers (CISC) and from uni–processor to multiprocessor architectures. **Chapter 2** presents two functional programming oriented VLSI architectures. **Chapter 3** looks at hardware support for programming languages to overcome memory limitations. Garbage collection (GC) helps to reclaim memory space that is no longer used by the program. In this chapter two garbage collectors are discussed; the first is for Lisp–like machines and the second for Prolog computers.

Computer architectures for knowledge oriented systems

Knowledge representation and manipulation tasks are frequently required in AI systems. These tasks have inherent parallelism which must be exploited in order to obtain reasonable execution times. **Chapter 4** deals with content–addressable memory (CAM) circuits. CAM circuits are useful for applications such as production systems and logic programming. The CAM implementations presented in this chapter illustrate the effective use of parallelism. In **Chapter 5**, two architectures for knowledge bases are described. The multiprocessor architectures are based on relational algebraic operations and semantic networks.

Neural network hardware implementations

In recent years many computer scientists have become interested in neural network models. Such models are believed to have a potential for new architectures for computing systems; such systems may be able to achieve human–like performance in some fields. **Chapter 6** looks at architectural implementations of neural networks which are based on the Hopfield model. **Chapter 7** presents several digital and analog circuits to implement these networks. The implementations reveal contrasting approaches to exploiting the VLSI capabilities and for overcoming the limitations imposed by this technology. **Chapter 8** gives some alternative designs for neural network computations. The computers presented here are not themselves based on a neural network model but they do, through more conventional conventional architectures provide high computational power for neural computing applications.

VLSI FOR ARTIFICIAL INTELLIGENCE

Chapter 1

PROLOG MACHINES

Prolog has been widely used over the past decade as a high–level language for artificial intelligence applications. As a consequence, much research has been carried out to develop high performance implementations of the language. These implementations range from sophisticated compilers and emulators to special–purpose firmware and hardware.

Most of the Prolog implementations are based on the Warren abstract machine (WAM) (Warren 1983). The abstract machine manipulates data in five basic areas: the *code area* contains the program; the *control area* contains the abstract machine registers; the *environment stack* contains information about backtracking and recursive procedure invocations; the *trail stack* contains references to conditionally bound variables; and the *heap* stores structures and values at execution time.

UNIPROCESSOR ARCHITECTURES FOR PROLOG

There is a wide range of uniprocessor architectures for direct execution of Prolog: from reduced instruction set computers (RISCs) to complex instruction set computers (CISCs) (Borriello *et al* 1987). In this chapter three different approaches are presented.

Eck §1.1 describes an approach for implementing basic operations in hardware and/or firmware in order to provide support for efficient use of a microprogrammable Prolog machine. In this approach the WAM instructions are implemented by means of elementary operations. Such operations are divided in three sets (namely, *data object primitives, explicit and implicit program control*). Each of the sets is analyzed and the requirements for specialized hardware devices are investigated. In this paper, two proposals for specialized hardware are discussed: automatically deferencing memory and additional comparison operations on tag patterns.

Civera *et al* §1.2 present an architecture that executes compiled Prolog. In order to obtain maximum performance of the hardware an evaluation of the environment, computational model and architecture is done. The resulting architecture is a 32–bit Harvard machine with completely horizontal code to allow the highest internal concurrency. This machine has three units: control, execution and bus interface.

A RISC approach for flat concurrent Prolog is presented by Ginosar and Harsat §1.3. CARMEL–1 is a 22–instruction processor that supports tag manipulation, type identification and dereferencing. The processor that can access memory twice every cycle may achieve 540 KLIPS executing *append.*

PARALLEL EXECUTION

Efforts towards parallel execution of Prolog have been concentrated in two major areas: AND–parallelism and OR–parallelism (Fagin and Despain 1987). AND–parallelism involves the simultaneous solution of subgoals in a clause. Since subgoals of a clause may share variables, variable binding conflicts may arise. OR–parallelism is the simultaneous unification of multiple clauses that share the same goal. During execution each variable may be bound to several values; therefore, a method to maintain a separate address space for each binding environment must be implemented on any OR–parallel system.

Reynolds and Delgado–Rannauro §1.4 study BRAVE, an OR–parallel dialect of Prolog, and propose VLSI hardware support for distributed access to data structures, memory management and task scheduling. Each of the processing nodes of the multiprocessor architecture has a processing element, message unit, a cache and a local memory.

References

Borriello, G., Cherenson, A. R., Danzig, P. B. and Nelson, M. N., "RISCs vs CISCs for Prolog: A Case Study," in *Int. Conf. on Architectural Support for Programming Languages and Operating Systems (ASPLOS II)*, pp. 136–145, October 1987.

Fagin, B. S. and Despain, A. M., "Performance Studies of a Parallel Prolog Architecture," in *the 14th Annual Int. Symp. on Computer Architecture*, pp. 108–116, June 1987.

Warren, D. H. D., "An Abstract Prolog Instruction Set," *Technical note 309, SRI International*. Melo Park, Calif., 1983

1.1 FROM LOW LEVEL SEMANTIC DESCRIPTION OF PROLOG TO INSTRUCTION SET AND VLSI DESIGN

Reinhard Eck

INTRODUCTION

Prolog machines presently under discussion for processing compiled Prolog programs are mainly based on two different principles of representing the programs:
- machine instructions for the host machine (e.g. the Warren Abstract Machine, WAM)
- a graph oriented model

The WAM code is a machine code with the von Neumann properties of sequential storage and sequential processing as well as conditional and unconditional branch instructions for structuring programs. Further properties of Prolog machines for the WAM code are dependent on the implementation of the instructions (e.g. the separation of code and data).

Machines for a graph oriented representation of programs do not have a fixed and well defined set of instructions. Instead, the programs are mapped onto data structures where the flow of control within the program is determined not by sequential storage and branch instructions but by explicit relations expressed by pointer structures. This model will not be discussed further in this article.

The WAM, originally designed as an intermediate code for the compilation of Prolog programs into some host machine language, proved to be best suited for implementation on the basis of a microprogrammable machine. The search for suitable base architectures is presently one of the main themes in the discussions about sequential Prolog machines and processors.

Current research and development efforts are aimed at the following two topics:
- design of hardware structures and devices for efficiently supporting the characteristic functions of WAM instruction set and implementation by microcode,
- tuning and optimization of the instruction set by modifying instructions or by adding new ones.

The following sections report a proposal for methodical steps to modifications of the machine language and the underlying hardware/ firmware level by observing and analysing data structures and operations employed in WAM implementation.

SEMANTIC LEVELS OF PROLOG PROGRAMS

In principle there are three levels of semantics of Prolog programs. On the upper level, the logical level, a logic program is the representation of a set of formulae of first order predicate logic specifying some relations of objects in a domain of discourse. In its meaning as a program this set of formulae is being implemented at a procedural level. After compilation of the program into the WAM code the program appears as a sequence of instructions of the WAM. The steps of the resolution principle are being implemented by these operations at the procedural level. The unification procedure itself, which is a part of the resolution procedure, is implemented at the next lower level which realizes the procedures of evaluating the parameters of the WAM instructions. For this level which is the level under consideration here there are three alternatives:
- the machine language level of some host machine on which the WAM code is mapped by a compiler,
- a direct execution hardware architecture for the WAM instructions,
- a firmware level which implements the WAM instructions and maps them onto an underlying hardware.

Only the last alternative will be discussed here. On this firmware level the semantics of a logic program can only be described by simple operators to be applied to objects residing in storage devices and registers. The functionality of those operators is offered by functional devices like arithmetic and logic units, sequencers and others.

Objects which have to be managed during program execution on that level are:
- representations of clause arguments - variables, constants, structured terms, list structures,
- the representation of the program, i.e. the sequence of WAM instructions,
- representations of information structures used for the implicit control of the program flow.

COMPLEX PROCEDURES IMPLEMENTED BY BASIC OPERATIONS

The problem addressed here is the search for elementary operations for efficient implantation of the WAM instructions. For the above mentioned objects there are three sets of operations. These are:
- operations on data structures which represent objects,
- operations for performing explicit program control,
- operations on data structures which are used for implicit program control (e.g. choicepoints).

The steps being described here for investigating such operations are firstly stating the functional requirements of the WAM implementation, secondly coping up with these requirements with a real implementation and thirdly deriving from that step a set of proposals for tuning the hardware/firmware levels implementation. In a fourth step one can, in addition, try to find out if there is some feedback to the development or modification of the instruction set of the WAM.

The steps will be discussed in the following sections starting with operations on data objects.

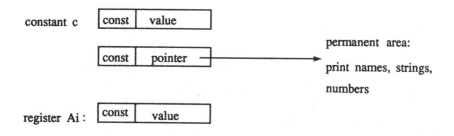

Figure 1a   Examples of constant structures

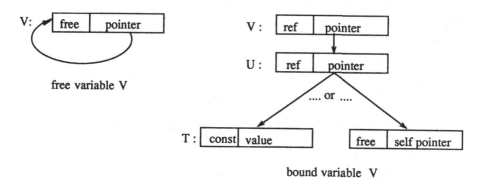

Figure 1b   Examples of structures of variables

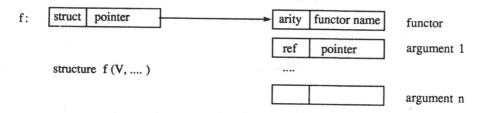

Figure 1c   Example of a structured term

Data Object Primitives

Data objects are the arguments of clause predicates and those structures which are derived from them dynamically by repeatedly applying the unification procedure during the execution of a program. Such structures are described by Kluzniak (1985) and can be shown as structures made out of memory locations, registers and their contents. Figure 1 shows typical structures for constants, variables and structured terms.

In order to find basic operations procedures on storage structures of data objects are inspected. For this sake the (semantic) operations on a storage structure (which possibly confines more than one object) are set against their implementation by microcode. The operations are distinguished according to the operational elements of the computer (storage, processor, microprogammable devices like ALUs, sequencers, register sets).

Starting at first with the distinction of memory access operations and "other" processing steps one can divide a procedure in two sorts of phases, memory access and processing phases internal to the processor. The registers used in the processor as source or sink of a memory access decouple storage transfer from processing steps on the data objects residing in the processor. Thus different phases of a (possibly complex) structure operation can be distinguished, where processing is performed only internal to the processor (see Figure 2).

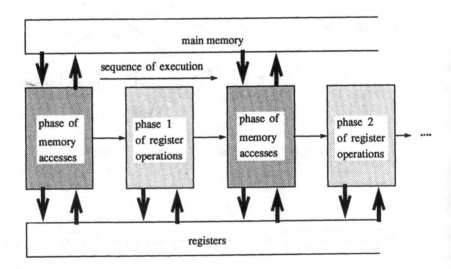

Figure 2 Phases of a Complex Operation

The correspondence between the semantics of a structure manipulation operation and a phase of processing steps internal to a processor is actually the correspondence between a behavioral specification and the implementation at the register transfer level.

This implementation in turn can be examined if it contains specific sequences of microinstructions or micro-operations which are worth combining to yield new operations or instructions for more efficient implementation of the required functionality. These ideas will be reinforced in the following sections.

In the course of resolving a goal clause in a logic program, unification has to be performed as a complex operation on the respective arguments of two predicates. By translation into WAM code unification is resolved into the corresponding instructions of parameter input (put_..., unify_...) procedure call (call ...) and parameter evaluation (get_..., unify_...) - see the following example instruction sequence (Figure 3).

```
          put_structure  F, Ai

          unify_variable Yj          # build a structured term
          .                          # on the heap
          call procedure

procedure:
          switch_on_term _,_,_,L1
          .
L1:       try_me_else   L2     # unification with a head
          get_structure F, Ai  # structure
          unify_variable Yk
          unify_value   Yk
          .
          proceed
```

Figure 3 Example of an instruction sequence

By checking the types of arguments the compiler is able to decide which part of the general unification procedure has to be performed and will select the appropriate sequence of WAM instructions. But in the worst case the whole unification procedure has to be activated, namely if a head variable cannot be specified in detail during compilation time. This procedure executed on two structures shall serve as an example for further considerations.

Given two terms represented by two bound variables TermA and TermB, which have to be unified. The meaning of an access operation to variable TermA is not the access to the location with the identifier TermA but to the object at the end of a possibly existing chain of references which is a constant, a free variable, a structure term or a list. Access to an object of type *ref* always intends access to the finally referenced object. A sequence of microinstructions for dereferencing actually implements an indirect object access which the memory devices currently in use cannot offer. Other types of memory access operations which have to cope with indirection are the evaluation of pointers in structure terms and list structures. The next section will discuss operations internal to the processor. One could define operations that are exclusive to the execution unit and operations that are exclusive to the control unit of the processor.

Operations of the execution unit of a processor. At this point the offer of functionality at the level of the programming language determines the requirements for efficient support by the firmware/hardware levels of a processor. Only requirements arising from the need to implement the

resolution principle, especially the unification procedure, are being considered here.
The analysis of the functionality demanded for by the WAM instructions and the analysis of the emulation of the WAM on a microprogrammable general purpose computer architecture, the Advanced Micro Devices Evaluation Board 29300, determine a set of requirements some of which are presented here.

Required alternatives for operand selection.
- read/write a whole register
- read/write the value field of a selected register
- read/write the tag field of a selected register

Required operations.
- comparison operations. The implementation of the unification procedure shows that frequent use is made of comparisons of the above listed operands. Tag combinations are compared for the sake of finding the appropriate unification routine for two objects. Value fields of two operands are compared for checking if they address the same object for instance.
- arithmetic and logic operations. The static analysis of the microcode for the emulation of the WAM on the above mentioned hardware shows that only 17 arithmetic and logic operations of the multitude of partially complex operations are currently being used. By appropriately providing for frequently used operands (numbers 0, 1, 2, 4) the increment and decrement operations could be left off. Nearly unused remain the logic operations AND and OR (see Table 1).

Table 1   List of the ALU operations used

operation	count	percentage
incr 1	96	3.23
incr 4	381	12.80
decr 1	44	1.48
decr 2	19	0.64
decr 4	549	18.45
shift	170	5.71
rotate	42	1.41
add	98	3.29
sub	195	6.55
xor	128	4.30
and	2	0.07
or	3	0.10
not	29	0.97
zero extend	154	5.17
extendF	72	2.42
passF	41	1.38
merge	953	32.02
total	2976	100.00

Operations of the control unit. The specification of the unification procedure shows that the control flow mainly depends on the combination of the tag values of two operands. A very efficient possibility for branching according to tag combinations is feasible with the multiway branch facility of the Am29331 sequencer. On the evaluation board, however, use of this facility can only be made by means of some intermediate steps which transfer the tags to be branched into the macroinstruction register. The direct access to the tag information for performing multiway branch operations is desirable.

Explicit Program Control

Explicit program control in programs in WAM code serves for selecting an appropriate clause alternative for a given procedure call (indexing instructions). By means of branch instructions depending on one argument (in general the first one) it is possible to select one clause from a sequence which fits the first argument in the sense of possible unifiability (called compatibility below). Starting at a label reached thus, the search continues incrementally by means of *try* instructions which create a choice point followed by a sequence of *get* and *unify* instructions and this way lead to the attempt to unify the arguments. That is why choice points would eventually be created even if the remaining arguments do not match according to their type.

The matching procedure can be refined and the preliminary creation of choice points avoided, if more information about the argument types could be obtained prior to the creation of a choice point. From the view point of the matching procedure choice points should not be treated if compatibility of arguments is not proved. A precondition for a matching procedure modified like that is that additional information about the types of the arguments involved is attached to the procedure call. A proposal for a facility like this is made in the section below on composing basic operations.

Implicit Program Control

In addition to explicit program control in processing compiled Prolog code one can find control mechanisms not expressed in the program representation: for the sake of efficiently backtracking in the search tree generated by a Prolog program and a given goal, reaching a choice point leads to the creation of a choice point data structure, which saves the machine state before entering one of the alternative branches.

Operations to be performed by the implementation levels are:
- create choice point data,
- reactivate the machine state from stored data,
- undo bindings which have been created since generation of the current choice point,
- force instruction pointer and instruction fetch mechanism to an alternative branch.

As discussed erlier in the section on *Data Object Primitives,* the complex task of implicit program control should be investigated for the possibility of extracting basic operations for its support in order to achieve further requirements for the microcoded implementation and microcodable circuitry.

COMPOSING BASIC OPERATIONS

The result of the steps described up to now is a set of requirements for operations which are useful for implementing WAM in microcode.
In this section we make two proposals which aim at the modification of

boards and circuits. They can also be understood as a step in the direction of specialisation of architectures for efficiently processing WAM code.
These proposed features are:
- an automatically dereferencing memory (interface),
- additional ALU comparison operations on tag patterns.

Selfdereferencing Memory. In the section on data object primitives the access to a variable of type reference is described as a scanning operation which holds at the last object in the reference chain. In the actual emulation of the WAM by the above mentioned microcodable board this operation is done by looping through a sequence of four microinstructions which evaluates the tag and value (address) fields of the memory locations in a reference chain. This takes four microinstruction cycles. The microprocedure is used in every instruction which is involved in unification (put, get, unify, switch instructions as well as the microroutines for unification and occur check). Provided its cost (time consumption for dereferencing in a set of benchmarks) justifies the development of special hardware support a memory interface with the following features will be considered:
- it detects if a word which is to be passed to the processor is a reference by checking the tag field,
- it places the pointer address of the reference immediately on the address bus to initiate a new fetch cycle,
- only the final object which is not a reference is passed to the processor,
- the interface signals its state to the processor,
- the function can be switched on/off by a microoperation.

By this means every reading access to a non-reference-object is to be initiated by one microoperation in the microcode. The status of the processor during a dereferencing action can be active until the next attempt is made to address the memory, or it can be idle.

Comparison operation on tags and tag patterns. The search for a candidate clause for unification can be supported by an additional comparison operation of the ALU.
A tag pattern is a string of tags. The basic tag values are:
- *var* for variables (the distinction between *reference* or *free* is not suitable here),
- *const* for constants (different types of const tags if they are distinguished in the data objects as well),
- *list* for lists,
- *struct* for structures.

Tag patterns are created by evaluating the argument types of goal and head predicates. The strings of two corresponding predicates have variable but identical length according to the number of arguments. The length is a parameter of the comparison instruction. The result of the comparison of two strings will for the sake of unification support not be "equal"/"not equal", but the indication of *compatibility* if every two corresponding tags within two strings are compatible according to the following Table 2. See also Figure 4 for the scheme of comparison.

Table 2 Compatibility of tags

tags	var	const	list	struct
var	c	c	c	c
const	c	c	-	-
list	c	-	c	-
struct	c	-	-	c

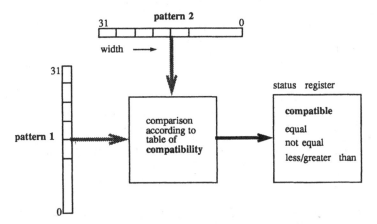

Figure 4 Comparison of tag patterns

As described earlier, this operation can be used during the implementation of the WAM indexing instructions in order to avoid premature creation of choice point frames in the control stack. In the program code the tag pattern information has to be carried along with the call and the *try* instructions. One proposal is shown in the following code sequence:

```
          pattern_call VCLS, label
          .
          .
label:
          switch_on_term L1,_,_,_
          .
          .
L1:    try_my_pattern_else SLCV, L2
```

In the storage representation of the modified instructions there has to exist some byte containing the tag count of the pattern and at least one word for the pattern itself. Creation of those instructions should not be a heavy burden for the compiler, because the WAM code in its unmodified form demands for evaluation of the argument types for selecting the right types of get, put, and unify instructions.

Contributions to Instruction Set Development

From the above sections one can find a relation between the division of single instructions into phases for analysing basic operations and the complexity of instruction sets. Three levels of complexity can be distinguished.

On the *first level* there are instruction sets like the WAM code. In most of all cases the implementations of the single instructions confine *more than one register transfer phase* and the interfacing memory access phases. One example is the unification instruction which is a vertically migrated algorithm. An instruction set like that could be named a *very complex instruction set*.

On the *second level* one can put instructions which confine not more than *one phase* of intra-processor actions, and which possess only one phase for operand fetch and one for storing results. This is the case with instruction sets of general purpose microprocessors currently in common use - *complex instruction sets*.

The *third* level contains instruction sets that can be achieved by refinement steps according to those treated in the section about data object primitives. The instructions can be divided in memory access instructions and those acting on internal registers. The result is an instruction set with *reduced complexity* which is comparable with the well known so called reduced instruction sets in terms of simplicity and the load/store scheme.

References

Advanced Micro Devices, *Am 29300 Family Handbook*, Advanced Micro Devices Sunnyvale CA 1985.

Advanced Micro Devices, *Am29300 Evaluation Board User's Guide*, Advanced Micro Devices Sunnyvale CA 1987.

Dobry, T.P. and Patt, Y.N. and Despain, A.M., " Design Decisions Influencing the Micoarchitecture for a Prolog Machine", *ACM SIG MICRO News Letter*, Vol. 15, No. 4, pp. 217 - 231, 1984.

Faign, B. et al. "Compiling Prolog Into Microcode: A Case Study Using the NCR/32-000", *ACM SIG MICRO News Letter*, Vol. 16, No. 4, pp. 79-88, 1984.

Flynn, M.J., "Towards Better Instruction Sets", *ACM SIG MICRO News Letter*, Vol. 14, No. 4, pp. 3 - 8, 1983.

Kluzniak, F. and Szpakowicz, S., *Prolog for Programmers*, London: Academic Press, 1985.

Kursawe, P., "How to Invent a Prolog Machine", *New Generation Computing*, Vol. 5, pp. 97 - 114, 1987.

Seitz, Ch.L., "Concurrent VLSI Architectures", *IEEE Transactions on Computers*, Vol. C-33, No. 12, pp. 1247 - 1265, Dec. 1984.

Stankovic, J.A., "The Types and Interactions of Vertical Migrations of Functions in a Multilevel Interpretive System", *IEEE Transactions on Computers*, Vol. C-30, No. 7, pp. 505 - 513, July 1981.

Stankovic, J. A., " Improving System Structure and its Effect on Vertical Migration", *Microprocessing and Microprogramming 8, pp. 203 - 218*, 1981.

Warren, D.H.D., "An Abstract Prolog Instruction Set", *Technical Note 309*, Artificial Intelligence Center, SRI International, 1983.

1.2 A 32 BIT PROCESSOR FOR COMPILED PROLOG

Pierluigi Civera, Dante Del Corso, Gianluca Piccinini
and Maurizio Zamboni

INTRODUCTION

The diffusion of the Logic Programming paradigm in many fields of the Artificial Intelligence requires the design and the implementation of new dedicated machines to improve the execution speed. Among the logic programming languages, Prolog is actually the most widely used. For this reason the realization of efficient Prolog machines represents an essential background for the development of AI techniques.

The work herein described deals with the different project phases of a VLSI implementation of a Prolog microprogrammed processor. These design steps can be summarized as follows:

−definition of the computational model which implements the language;
−evaluation of the computational model;
−definition of the architecture from the model;
−evaluation of the architectural features;
−definition of the microarchitecture of the processor;
−translation of the high level execution algorithm into the microcode.

The abstract model chosen is based on the efficient execution of a compiled code. The instructions belong to the Warren Instruction Set and are directly executed by the processor. The mapping of the abstract machine on the processor architecture is performed with the aid of evaluation and synthesis programs. The resulting microarchitecture has been simulated and the processor itself is now being implemented on silicon.

The Prolog processor is conceived as a dedicated 32 bit coprocessor working with a general purpose CPU. The main difference between this solution and the other dedicated coprocessors, such as floating point units, consists of its capability to fetch segments of code stored in memory autonomously. The coprocessor executes the code directly and interfaces with a standard CPU during the initialisation phase and when returning the results; moreover the CPU is called whenever a built-in instruction is not directly executable by the coprocessor. To increase the performance a prefetching unit is added to the architecture and placed in front of the coprocessor.

THE COMPILED EXECUTION OF PROLOG

The execution of Prolog programs requires the definition of a computational model which implements the primitives of the language. Two different computational models have been considered: one for the interpreted and one for the compiled execution.

The interpreted execution allows a simpler implementation of the extralogical predicates, even if the behavior of the computational model in terms of memory accesses and internal operations is less efficient with respect to the compiled execution (Civera *et al* 1987a,b). The highest performance is obtained with the compiled version discussed in the paper.

The computational model chosen for the compiled execution was developed by Warren (1983). The Warren Abstract Machine (WAM) is based on five data structures:

–*the Environment Stack*
–*the Heap*
–*the Trail*
–*the PDL*
–*the A_x Register file*

and the following registers:

–*P: Program pointer register*
–*CP:Continuation Pointer register*
–*E: Environment stack pointer*
–*B: Backtracking choice point register*
–*H: Heap write register*
–*HB: Heap Backtracking pointer register*
–*S: Structure pointer register*
–*TR: Trail stack pointer*

The WAM introduces some optimizations on the management of the variables involved in the predicates unification. The model considers permanent and temporary variables and only permanent ones are stored in the Environment Stack; temporary variables are kept in the register file A_x of the processor and are passed to the called procedure as parameters.

The WAM model handles structured data and lists using the Non Structure Sharing (NSS) technique. Some Warren instructions manage Prolog non-determinism creating, modifying and destroying the choice points inside the environment stack with a technique widely applied in the interpreted execution.

THE EVALUATION ENVIRONMENT

The implementation of a dedicated Prolog processor requires the definition of a design methodology which involves both the computational model and the architectural aspects. Good performance can be obtained only if the execution algorithm is well matched to the architecture, using the hardware resources in the best way. This implies a good knowledge of the effects of the architectural features on the processor performance. To collect this information an evaluation environment has been developed. It is built on a high level simulator which executes the Warren code.

The intermediate code is produced by the PLM compiler (Van Roy 1984, Touati *et al* 1987). This compiler, written in Prolog and developed at UCB by Peter van Roy since 1984, uses an instruction set very close to the Warren one but some instructions for the unification of the lists are added implementing the car/cdr notation on structured terms.

A "flattening" program expands the high level simulator considering the physical architecture. The output of the program is still a simulator but at an "atomic" level, where each operation corresponds to a hardware microoperation.

During the execution of the benchmark programs a detailed trace file is generated. This file contains all the information describing completely the computational state in terms of both memory references and operations on the WAM registers. The trace files are analyzed by two different tools:

   −a statistical evaluator of the elementary operations;
   −an architectural evaluator.

The former allows a statistical characterization of the computational model, giving the distribution of memory accesses and the internal operations performed on the WAM registers.

The latter (Architectural Prolog Evaluator APE) is a more sophisticated tool written in Prolog which analyzes the trace files considering the data dependencies and the architectural constraints (Civera *et al* 1988a). This program can evaluate the real performance and many other factors starting from the description of a machine architecture.

THE COMPUTATIONAL MODEL EVALUATION

The first step to define the processor and the overall system architecture is the analysis of the computational model.

A first class of data are derived from the memory behavior of the model. Effective decisions can be taken from these data about the memory interface structure of the processor itself (Tick 1987).

A second class of measurements concerns the distribution of internal operations on the abstract machine registers. These values are important in order to decide about the physical implementation of the abstract machine registers.

The results are derived from the execution traces of some benchmark programs such as:

- *nrev:* naive reverse of a list of 30 elements,
- *qsort:* quick sort of a list of 50 numbers,
- *color:* a color mapping algorithm,
- *sieve:* Eratosthenes' algorithm for the prime numbers.

The memory operations have been measured in terms of:

- data structure accessed (Code/Data, Heap/Environment Stack,...),
- distribution of the accesses (locality of the accesses),
- memory operations for Logical Inference.

The distribution of the data memory accesses for the *"qsort"* benchmark is reported in Figure 1. The represented values are expressed in number of read/write accesses per Logical

Inference. The code accesses have been evaluated considering a variable length code.

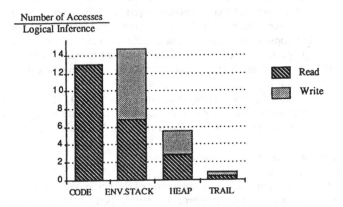

Figure 1 Distribution of the memory accesses for the *qsort* benchmark

The code memory accesses represent about 40% of the total number of memory operations and this suggests a Harvard architecture for the processor. Considering the data memory the environment stack involves about 60% of data accesses.

Trace data about memory operations point out that a performance limiting factor of the WAM is the high memory bandwidth required per Logical Inference.

In Figure 2a the overall performance is shown as a function of the memory access time and the memory occupation factor, considering a single port for data and code memory. The memory occupation factor (M) represents the ratio between the memory accesses and the total number of processor cycles. The curve with unity memory occupation factor is the theoretical limit of the model for a single bus architecture. In practice data dependencies reduce the memory occupation factor. In our implementation M is about 0.4. Architectural solutions such as pipelining, interleaved accesses, caching, separated code and data memory improve the performances and raise the theoretical limit.

An architecture with two separated busses (for data and code) was considered. Data memory accesses, being more frequent than code references, define the limit. Figure 2b shows the curves for this solution.

Similar evaluations and analyses have been performed for the internal operations of the WAM. This step is fundamental to raise the internal limits imposed by the computational model. The internal cost depends on the number of operations, their execution sequence and the time spent for the execution. Therefore the objects of the abstract machine have to be characterized in terms of their functionality and connectivity. The information which can be obtained is divided in two classes:

- frequency of logical/arithmetic operations on registers;
- transfer operations among the registers and among registers and the external memory interface.

Figure 3 summarizes some of these values. They have been obtained considering the dynamic traces and represent the statistical distributions of the internal operations.

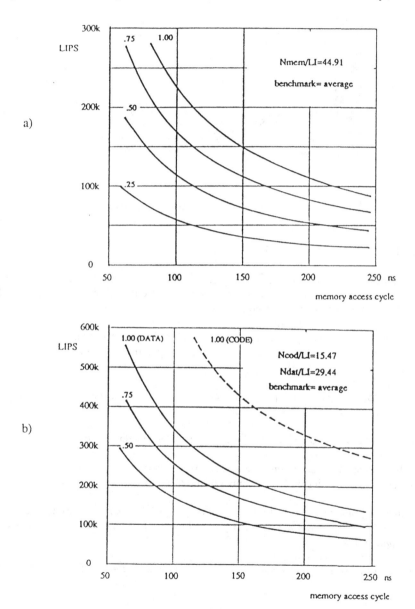

Figure 2 Limiting curves for different memory occupation factors *vs.* memory cycle time. The number of memory accesses per L.I. is the averaged value measured on *Color, Cmos, Hanoi, Mobius, Nrev* and *Qsort* benchmarks.

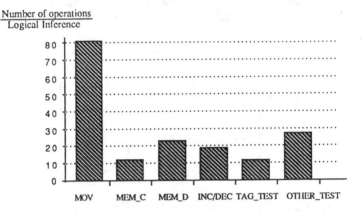

Figure 3 Distribution of the internal operations per L.I. for *qsort* benchmark

THE ARCHITECTURAL EVALUATION

The mapping of the abstract machine onto a physical architecture requires the evaluation of the overall performance with respect to the different architectural choices (Dobry 1987, Habata *et al* 1987). As previously introduced the APE program has been designed to evaluate the performance from the trace data after consulting an architectural description file. Being more involved in the processor design, the program has been used in the definition of the processor microarchitecture, based on the assumption of a microprogrammed machine.

Two sets of measurements related to the processor performance and to the bandwidth occupation factor have been collected:
- the "degree of compaction" of the program traces (dynamic) and the microcode (static);
- the sensitivity of the architecture to some design features.

K is the degree of compaction which can be evaluated in two different conditions:

K_d = (number of executed trace microorders)/(number of cycles performed)
K_s = (number of control microorders)/(number of microcoded words).

These parameters point out the capability of the considered microarchitecture to execute more than one microorder per machine cycle. Higher values of these parameters increase the bandwidth occupation factor.

The sensitivity of a given architecture to a parameter has been defined as follows:

$$S = \frac{dK_{s/d}}{K_{0\ s/d}}$$

	NREV (K_{dyn})	QSORT (K_{dyn})	MICROCODE (K_{stat})
S_{busA}	-13.12	-11.7	-12.7
S_{busC}	+4.52	+4.87	+2.35
S_{ms}	-3.2	-4.39	-7.54
S_{nobrd}	-28.5	-21.46	-26.4

Figure 4 Sensitivity factors for different architectural parameters
S : Sensitivity to an architectural parameter (%)
K: Degree of compaction (s/d)
busA: Internal bus A
busC: Additional third internal bus
ms: Master slave capability for MAR and MDR registers
nobrd: No broadcast transfer capability for the internal busses

Figure 4 indicates some values evaluated on the implemented architecture by varying one architectural parameter at a time and measuring the degree of compaction obtained. The obtained values allow the designer to choose different architectural features on quantitative basis.
The methodology has been used to define the following design decisions:
- number of internal busses,
- horizontality of the microcode,
- memory interface mechanism.
As an example Figure 5 points out the degree of compaction (which can be seen as a speed up factor) of the execution (dynamic trace) of the nrev30 considering 1,2 and 3 internal busses microarchitecture.
The complete results of this analysis and their architectural implications are reported in (Civera *et al* 1987b).

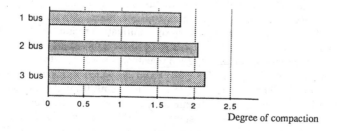

Figure 5 Degree of compaction of the execution of the *Nrev30* considering 1,2 and 3 internal busses microarchitecture.

THE PROCESSOR ARCHITECTURE

The resulting processor architecture, shown in Figure 6, is based on a Harvard structure: one port for the data and one port for the code. The processor manages the data memory directly and uses an external unit which prefetchs and aligns the code read from the program memory. The internal architecture is based on a horizontal microprogrammed machine: the defined data path implements the highest degree of concurrency inside the same microword.

The processor is divided into three main parts: control unit, execution unit and bus interface. For the control unit the simulations showed that a good level of concurrency can be obtained, so a completely horizontal code was chosen. A word of code is divided in two parts: the first defines the control signals to be sent to the execution unit and the second is related to the next microaddress selection and contains both the controls sent to the microsequencer and the address seed. The whole microword is pipelined inside a register called the Micro Instruction Register (MIR).

The sequencer is designed so that the computation of the next address is performed using the address seed. This avoids the use of an incrementer and permits the allocation of the code in non sequential addresses.

The address seed is used also during branches as a base address. The algorithm requires many different types of tests, grouped into two different classes: binary and multibranch tests. Multibranch tests are data type tests (tag) and can be solved either with a chain of binary tests or with a faster single cycle multiple branch. The selected implementation of binary branches evaluates the branch address using a status signal as the least significant bit (LSB) of the address seed. This technique requires that the two branch addresses are consecutive; the address seed must also be an even address.

In case of a multiple branch the technique is extended to others LSBs of the address seed (two for 4-ways etc.); the N-way branch is therefore implemented in a single clock cycle; notice that the address seed must have the two LSB forced to 0. This technique permits resolution of single and multiple branches in a single cycle but needs greater attention to allocate the code to avoid wasting of microcode memory.

The algorithm structure and the need for reduction of the code required the identification of parts of the code which could be transformed into subroutines; it seemed worthwhile to implement a microsubroutine mechanism so an analysis of the occurrence of the calls has been made. The calls are not nested so it is not necessary to provide a mechanism of stacking of the return addresses.

On the other hand the stacking mechanism is needed by the unification routine, since this is not a pure sequence of microinstructions but has also internal recursive calls. In this case it is mandatory to store the status of the computation somewhere to permit the correct execution of the nested unification calls. The analysis of the unification routine showed that the recursion starts at only four points of the routine and each call is accompanied by the saving of two internal registers. Since two tag bits of the saved registers are not used, they are replaced by a 2 bit code representing the calling point of the routine; in this way it is not necessary to store the return address in a stack. The correct return address can be rebuilt from these two bits.

The execution unit consists of different objects (registers, ALU, counters) needed by the algorithm, connected by two busses. The study of the execution activity suggested a two bus solution for the data path as a good trade-off between speed and complexity. The same analysis provided an opportunity to decentralize increments, decrements, zero tests operations directly onto registers that have been transformed into up/down counters. Ten point-to-point links have been implemented among the registers to increase the concurrency of the microoperations.

Some complex but often used tests are solved in the execution unit with dedicated magnitude comparators. The tests can be solved in a single clock cycle and the output to the control unit is a single status signal for each type of complex test.

Great care has been used in the design of the bus interfaces since the interaction with the memory may represent the bottleneck of the project. The Data memory interface is connected via two registers which share an external multiplexed bus: Memory Address Register (MAR) and Memory Data Register (MDR).

At the beginning of every memory cycle the content of the MAR is placed on the bus; in a second phase the MDR places or collects data to/from the bus, according to the type of memory operation (write or read). In several cases the address of the MAR must be incremented after the memory access and the old MDR content must be available during the first phase of the internal machine cycle.

These considerations lead to the implementation of MAR and MDR as master-slave registers: even if the MAR or MDR are updated, their outputs to the address/data bus are modified only at the end of the internal machine cycle.

The interface toward the Code memory is simpler since the data transfers do not require any addressing mechanism. The interface is therefore reduced to an input-output port (via the PREF register) that can receive data from the prefetcher (the Warren instruction code) or send data to the prefetcher (the content of the Program counter register during prefetching re-queueing).

The interface also signals to the prefetcher when a new code can be loaded in the processor (code buffering is available inside the processor) and if it is possible to fetch new codes from the memory.

The main registers which form the data path are:

A_x-register file;
I-code address of A_x;
M-frame address of A_x;
MP-active register counter;
N-permanent variable displacement register;
NP- number of permanent variables used in the current call;
E-Environment stack pointer register;
H-Heap pointer register;
S-Structure pointer register;
HB-Heap backtracking pointer register;
B-Backtracking register;
CP-Continuation Pointer register;
TR-Trail register;
PDL- push down list pointer register;
PREF-prefetching register;
F1,F2,F3-fetching registers;
C-constant register;
R1,R2-temporary registers;

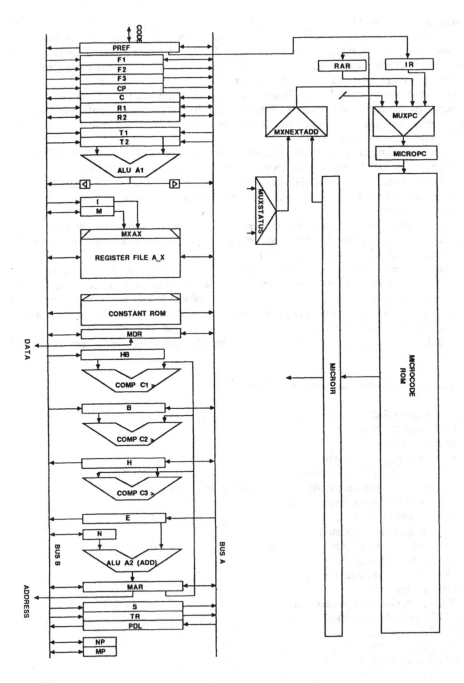

Figure 6 Block diagram of the processor

THE PREFETCHER STRUCTURE

The variable length code allows savings of memory space and reduces the memory bandwidth required by the Prolog processor itself.

The alignment of the code in a fixed format is a quite complex task and the implementation on the same chip as the Prolog processor is infeasible. So the partition of the processor into two chips simplifies the area and testability problems for the VLSI prototype. Moreover, if one chip is dedicated to the fetch operations it can be extended to a prefetcher role.

The introduction of the prefetcher implies a careful study of the interface between the prefetcher and the Prolog processor to achieve good performance; the aligned code is directly sent to the processor via a dedicated 32-bit port.

The alignment phase requires a manipulation of the original byte code to match the needs of the processor; the most significant operation is the expansion of the opcode from 8 to 10 bits. This translation transforms the opcode into the physical address of the microcode. As a consequence the Prolog processor does not decode the instruction; it simply loads the instruction opcode into the microaddress register thus decreasing the latency related to the decoding of the instruction. Such a choice also makes the testing of sections of the microcode easier.

The P register contains the address of the code memory location to be executed by the processor. It is updated internally during fetch operations and by the processor in the case of instruction requeueing. The P register is placed in the prefetching unit. This choice creates some problems in the management of the P content between the prefetcher and the Prolog processor. Three kinds of operation require a more complex management of the P register:

- the FAIL/GO_TO instruction,
- the CALL instruction,
- the TRY and RETRY instructions.

These instructions require the modification of the P or CP registers as well as the discharging of the prefetcher queue.

In summary, the main tasks performed by the prefetching unit are:
- alignment of the variable length code into a fixed format,
- translation of the opcode into the physical address of the microcode,
- management of the P register in the instructions which require to know its value or which change its content,
- managing of a prefetching buffer.

THE VLSI DESIGN

The design goal is to integrate the Prolog processor in a single integrated circuit. Complexity and power considerations suggest the CMOS technology as the only candidate for the integration. To obtain a fast and reliable development, the silicon compilation approach (Cheng *et al* 1988) has been pursued with the added benefit of an easy redefinition of the design. The diagram depicted in Figure 6 is composed of three main parts: the control store, the execution unit and the sequencer (including the bus interface). Each part is compiled using different GENESIL elements; the control store is implemented with the ROM block, the execution unit using the data path structures and the sequencer with standard cells.

Attention has been paid in the floor plan step to avoid long block interconnections and to reduce routing areas (Figure 7).

The complete data path is composed of 35,525 transistors; using a 1.5 microns N-well CMOS technology the resulting size is 2.736 mm by 17.520 mm. The considerable length of the data path is due to the large number of registers, ALU and counters and is impractical for a direct implementation. In the final implementation the data path is broken into two segments. The ROM block is placed on the top of the two data path segments with the sequencer and the bus interface on its sides to reduce the interconnection length. The content of the microcode ROM is automatically generated by the APE program from the description of the final layout. The APE program swaps the position of the control bits (ROM columns) accordingly.

A LSSD implementation of the microinstruction register was introduced to improve the IC testability. The LSSD microinstruction register, with the code and the data bus interfaces are arranged during the test mode to allow an easy inspection of the microcode. Furthermore the microinstruction register can be serially loaded with external test microinstructions. This technique allows a simple stimulation of the whole execution unit.

The list of the main blocks of the microarchitecture with their dimension and complexity is reported in Table 1. The estimated die area is 120.4 mm^2 including bonding pads. The chip will be housed in a 144 pin grid array package; the free pins are used as additional test points.

The first silicon is expected at the end of the year. The timing analysis and simulations are now considered. Critical paths are checked; the slowest element in the data path, the ALU, gives 43.9 ns as a worst case propagation delay, which is still consistent with the target speed of a 10 MHz clock operation.

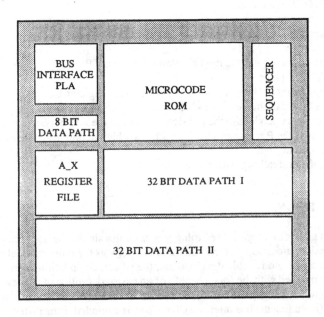

Figure 7 Floor plan of the Prolog processor

Table 1 Block size estimates

Data Path Block	Block Height (μm)	Block Width (μm)	No of Transistors
Pref_C_R1_R2	2506	2389	5349
F1_F2_F3_CP	2228	708	1493
Alu_T1_T2_Acc	2456	1385	4082
Mdr	2506	1218	2732
Mar	2736	1855	4490
C11_HB	2469	800	2025
C12_B	2469	1210	3081
C13_H_S	2469	2202	5341
Adder_E_N	2456	908	2474
Tr_Pdl	2458	1901	4458
A_x	2736	2952	n/a
Micro_code_Rom	2378	6203	n/a

CONCLUSIONS

The paper refers to the implementation studies and the design of a VLSI Prolog Processor for an Advanced Workstation for Artificial Intelligence (AWAI). The project is supported by the Italian Research Council (CNR) in the "Progetto Finalizzato Materiali e Dispositivi allo Stato Solido" (PF- MADESS) "Microstrutture VLSI".

From an accurate analysis of the computational model the architecture has been obtained and evaluated. The processor layout has been designed using the silicon compilation technique and the first silicon is expected at the end of 1988. A second integrated circuit which will work with the Prolog processor as a Prefetching Unit is under development.

The future work includes a plan to extend the processor functionality toward the OR-parallel execution environments.

REFERENCES

Cheng E.K. and Mazor S.,"The Genesil Silicon Compiler", in *The Silicon Compilation*, Addison-Wesley pp.361-405 1988

Civera P.L., Maddaleno F., Piccinini G.L.and Zamboni M., "An experimental VLSI Prolog Interpreter: Preliminary Measurements and Results", in *Proc. 14th Annual International Symposium on Computer Architecture* pp. 117-126 1987a

Civera P.L., Del Corso D., Piccinini G.L. and Zamboni M., "AWAI Prolog Co-processor: Analysis results and architecture definition", *Tech. Rep. DE-0987*, Dip. di Elettr.,Politecnico di Torino, Italy 1987b

Civera P.L., Piccinini G.L. and Zamboni M., "Using Prolog as Computer Architecture Description and Synthesis Language", submitted to *IFIP conference* September 1988a

Civera P.L., Piccinini G.L. and Zamboni M., "A VLSI Prolog co- processor: Implementation Studies", to be published in *IEEE MICRO* October 1988b

Dobry T.,"A High Performance Architecture for Prolog", *Ph.D. Thesis* Computer Science Division TR-UCB/CS 87/352 University of California Berkeley 1987

Habata S.,Nakazaki R.,Konagaya A.,Atarashi A. and Umemura M., "Co-operative High Performance Sequential Inference Machine : CHI" , in *Proc. International Conference on Computer Design: VLSI in Computers & Processors* pp. 601-604 1987

Tick E.,"Studies in Prolog Architecture",*Technical Report No.CSL-TR-87-329* Computer Systems Laboratory Stanford University June 1987

Touati H. and Despain A. "An Empirical Study of the Warren Abstract Machine" in IEEE proc. Symposium on Logic Programming September 1987

Van Roy P., "A Prolog Compiler for the PLM" Computer Science Division, TR UCB/CS 84/263, University of California Berkeley November 1984

Warren D.H.H., " An abstract Prolog Instruction Set" Technical Note 309, SRI 1983

1.3 CARMEL–1: A VLSI ARCHITECTURE FOR FLAT CONCURRENT PROLOG

Ran Ginosar and Arie Harsat

INTRODUCTION

Flat Concurrent Prolog (FCP) is a process-oriented, OR-nondeterministic parallel logic programming language (Shapiro 1986). It is intended as a "natural" tool for programming highly parallel message-based computers. FCP is the language of our multicomputer CARMEL (Computer ARchitecture for Multiprocessing Execution of Logic programs). In this paper we describe CARMEL-1, a high performance uniprocessor component of that parallel architecture.

We have carried out an architecture-oriented execution analysis of FCP (Harsat 1987). It employs a novel structured methodology for the optimal design of CARMEL-1. This methodology is based on, and extends, the well known RISC concept (Katevenis 1984, Gimarc and Milutinovic 1987) by suiting it for logic programming languages like Prolog and FCP. The analysis includes, a definition of a four-level language hierarchy: FCP, FCP abstract machine (FAM), and the machine (simple, RISC-like) instruction set level. The fourth, is a *novel* intermediate level of primitives (PL), which is defined to focus the analysis at the desired architectural level. Basically, PL is inserted in the gap between FAM and the lowest machine level. However, it includes components from all three levels mentioned above. PL forms the basis for CARMEL-1 instruction set.

In our analysis we have employed an experimental early software prototype of FCP (Houri and Shapiro 1986). As such, the architecture of CARMEL-1 necessarily carries some of the disadvantages and inefficiencies of the original software environment. CARMEL-1 achieves 540 KLIPS executing *append*. CARMEL-2, the successor of CARMEL-1, is currently being designed. Its architectural enhancements over CARMEL-1 are briefly described in the last section. In addition, we investigate the incorporation of the CARMEL-2 uniprocessor in a full multiprocessor. We expect this investigation to further influence the uniprocessor architecture.

In the following sections we survey various characteristics of FCP and the results of its execution analysis; then the system architecture is described, followed by data types, instruction set, pipeline, data path, and performance evaluation. In the last section we discuss the future work and conclude the paper.

FCP AND ITS EXECUTION ANALYSIS

In our implementation, FCP is compiled into a special serial *FCP abstract machine* (FAM) (Houri and Shapiro 1986). The FAM is similar to the well known WAM (Warren 1983), and differs mainly in being multiprocessing oriented. Also, unlike WAM, no environment stack is maintained and no backtracking takes place. In addition, FCP introduces a synchronization mechanism of read-only shared variables.

Each process is described by a record in memory, which identifies its code and data. A ready ("active") queue of processes is maintained. Processes block when accessing noninstantiated read-only variables. A semaphore-like mechanism wakes up the suspended process when the variable gets instantiated.

Data are globally shared, except for a small number (eight) of each process' arguments. The process state consists of a mere four FAM registers (CP: current process pointer; QF: queue front, i.e. next active process pointer; PC: program counter; A: argument pointer). Process switches are very frequent. When processes switch, two registers are restored from memory (QF, PC) while the other two, CP and A, are internally computed. In other words, the processes are "light weight," in the sense that they switch often and fast.

The FAM architecture includes 16 special purpose registers, and several data structures for process and dynamic memory management. Data memory is dynamically allocated within a single heap. FCP, as a non-procedural language, supports no user subroutines. Subroutines are used only to implement system services and guards. They have no local variables, and very few arguments are passed. Hence all calls are to predetermined locations. Most branches are also directed at absolute addresses; this is due to the fact that FCP is free of the concept of programmable control flow: there is no GOTO, etc.

The execution profile analysis presented by Harsat (1987) and by Ginosar and Harsat (1987) reveals the following findings. Dereference operations take 22% of the total execution time. Various pointer manipulations account for about 20% of the time, beyond dereferencing. Type identification consumes over 15%. Call and return overhead of system predicates (called *guards*), takes at about 9% of the time. We have found that FCP programs demonstrate a characteristic behavior, independent of parameters like the type of computation, the size of the program, run-time memory requirements, and others. In addition, garbage collection (which is a system service rather than a FCP inherent activity), does not affect significantly the characteristic behavior. These findings are the basis for the FCP support in CARMEL-1.

CARMEL-1 SYSTEM ARCHITECTURE

Figure 1 describes the system architecture of CARMEL-1. Data memory is separate from instruction memory. Both memories may be accessed within a single machine cycle. CARMEL-1 places the instruction and data addresses on the address bus one after the other, in the beginning of a cycle, and each address is captured by the corresponding address latch. At the beginning of the following cycle the instruction is fetched, followed by the required data word, on the data bus.

By combining two accesses during the same cycle we take advantage of the relatively long memory access time. A different design with only one type of access per cycle would not have rendered shorter cycle times. The Jump and Call detector unit is used for fast decoding and target prefetch of unconditional Jumps and Calls, and is explained below.

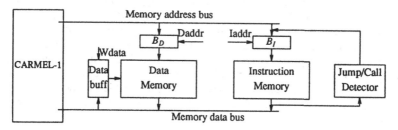

Figure 1 CARMEL-1 System Architecture

Data memory is partitioned into seven segments. Garbage is collected only in the main data structure, the *heap*, when it becomes full. A double-buffering scheme is employed. Memory partitioning keeps six relatively small data structures outside the heap, in specially managed areas, to reduce the rate at which the heap fills up, and thus to reduce the overhead of heap copying.

DATA FORMATS AND TYPES

There are nine types of data arguments in FCP (see Figure 2).

argument type	tag	# of data bits
1. *integer*	11	30
2. *list-integer*	10	"
3. *reference*	0111	28
4. *read-only reference*	0110	"
5. *string*	0101	"
6. *tuple*	0100	"
7. *list-reference*	0011	"
8. *list-ro-reference*	0010	"
9. *variable*	0001	"

Figure 2 Data Formats and Types

Each argument is accompanied by a tag. Two types, *integer* and *list-integer,* contain real data. The remaining seven (*variable, tuple, reference, read-only reference, list-reference, list read-only reference* and *string*) contain pointers. Thus, four bits are sufficient for the tag. Our tag allocation allows 30-bit integers and 28-bit addresses.

CARMEL-1 INSTRUCTION SET

The CARMEL-1 instruction set is defined according to the execution analysis carried in (Harsat 1987). It also reflects pipeline dependencies, as described in the following section. The instructions are shown in Table 1. In the table, we use *tag()* and *value()* to represent the tag separation hardware. *Tag()* returns the tag part of an argument. *Value()* returns a 30-bit integer, if the tag is *integer* or *list_integer*, and a 28-bit value sign-extended to 30-bit otherwise. 'll' represents concatenation. If the ALU output is stored in a register, it is concatenated with the tag of the first argument, to produce a 32-bit result. In all other cases, the ALU output is a calculated address, which is forwarded to either the PC, or the address bus, or the data bus. In these cases, the ALU 30-bit address is truncated to 28-bit.

Table 1 CARMEL-1 Instruction Set

Group	CARMEL	Instructions	Function
Arith-metic	ADD	R_s, S_2, R_d	$R_d \leftarrow [value(R_s) + value(S_2)] \parallel tag(R_s)$, set CC
	SUB	R_s, S_2, R_d	$R_d \leftarrow [value(R_s) - value(S_2)] \parallel tag(R_s)$, set CC
	XOR	R_s, S_2, R_d	$R_d \leftarrow [value(R_s) \oplus value(S_2)] \parallel tag(R_s)$, set CC
	AND	R_s, S_2, R_d	$R_d \leftarrow [value(R_s) \& value(S_2)] \parallel tag(R_s)$, set CC
	OR	R_s, S_2, R_d	$R_d \leftarrow [value(R_s) \mid value(S_2)] \parallel tag(R_s)$, set CC
Shift	SLL	R_s, R_d	$R_d \leftarrow logic\_shift\_left(value(R_s)) \parallel tag(R_s)$, set CC
	SRL	R_s, R_d	$R_d \leftarrow logic\_shift\_right(value(R_s)) \parallel tag(R_s)$, set CC
	SRA	R_s, R_d	$R_d \leftarrow arith\_shift\_right(value(R_s)) \parallel tag(R_s)$, set CC
Load and Store	LOAD	$S_2(R_x), R_d$	$R_d \leftarrow M[value(R_x) + value(S_2)]$
	LOADr	Y, R_d	$R_d \leftarrow M[PC + Y]$
	STORE	$Imm<16>(R_x), R_m$	$M[value(R_x) + Imm<16>] \leftarrow R_m$
	STOREr	Y, R_m	$M[PC + Y] \leftarrow R_m$
Flow Control	JMP	Address	$PC \leftarrow Address$
	JC	$COND, S_2(R_x)$	$if(COND)\{ PC \leftarrow value(R_x) + value(S_2) \}$
	JCr	$COND, Y$	$if(COND)\{ PC \leftarrow PC + Y \}$
	CALL	Address	$push\_stack(PC)$ then $PC \leftarrow Address$
	TRAP	$COND, DI, S_2(R_x)$	$if(COND)\{push\_stack(PC)$ $\quad\quad then\ PC \leftarrow value(R_x) + value(S_2)$ $\quad\quad if(DI)\ disable\ interrupts\}$
	RET	$COND, EI$	$if(COND)\{pop\_stack(PC),$ $\quad\quad if(EI)\ enable\ interrupts\}$
FCP Special	InsTag	R_s, Tag, R_d	$R_d \leftarrow value(R_s) \parallel Tag$
	IfTag	R_s, Tag	$if(tag(R_s) \neq Tag)\ skip$
	BRonTag	R_s	9-way branch by $tag(R_s)$ The 9 addresses are in $PC+1, PC+2,..., PC+9$
	Deref	Rd_2, F, R_s, Rd_1	R_s holds initial pointer $if(F)\ set/reset\ R\ (read-only\ flag)$ when read-only-ref is found/not found $Rd_1 \leftarrow$ final pointer $Rd_2 \leftarrow$ dereferenced value

S_2 is either a second register or a 16-bit 2's complement immediate operand. Y is a 22-bit 2's complement displacement for PC-relative addressing mode. The Condition Code (CC) register contains five flags: N, Z, V, C and R for negative, zero, overflow, carry and read-only, respectively. *Tag()*, *value()* and 'll' are explained in the text.

The instruction set includes only 22 instructions. The *shift* instructions provide only a single position shift. There are no explicit multiple bit shifts in FCP, and the implementation does not require such shifts either. Thus, the data path does not include a barrel shifter (Katevenis 1984). The single bit shifts of CARMEL-1 provide only the basic support for multiplications and divisions.

Special Instructions

The whole instruction set was designed carefully to efficiently support FCP. While most instructions in the set resemble those appearing in other processors (Gimarc and Milutinovic 1987), their specific implementation is unique. Four instructions are special, and are presented as a result of the peculiarities of FCP: *Insert-Tag (InsTag), Skip-If-Tag (IfTag), 9-Way-Branch-on-Tag (BRonTag),* and *Dereference (Deref)*.

InsTag inserts a given immediate tag into the four msb's of the specified register, thus making a legal argument out of the given tag and the value stored in that register. The opposite operation of value extraction is unnecessary: The value may be used as an integer or as an address (pointer). In the former case, only the 30 bits making the integer are routed into the ALU; in the latter case, only the 28 bits of the address are actually used. Thus, in both cases the tag bits are ignored.

Tag identification is carried out by the 9-way Branch-on-Tag *(BRonTag)* instruction. The instruction spans ten memory words, one for the opcode and the operand, and nine for alternative JMP instructions (one per each tag). BRonTag takes three cycles altogether. During the first cycle the new PC value is determined (one of PC+1,...,PC+9) according to the argument tag. During the second cycle the appropriate JMP instruction is fetched. That JMP is executed during the third cycle. Note that exploiting the regular JMP instruction accounts for two advantages. First, the processor design is regularized: there is no need to devise a special case of jumps, and the resulting controller is simpler. Second, the Jmp/Call detector is utilized automatically, without having to make it sensitive to BRonTag as well.

In many cases, FCP requires to identify a single specific tag, rather than sort out nine cases. The Skip-if-Tag *(IfTag)* instruction is used in such cases to save space. The tag of the argument is compared with the immediate operand included within the instruction. The next instruction is skipped if not equal. When the skip is not taken, IfTag consumes only a single cycle. Otherwise, one extra cycle is spent waiting for the next instruction to be fetched.

Deref is the instruction for linked list dereference. The number of cycles needed for completion of *deref* is L+1, when L is the length of the reference chain. Upon completion, in case of a pointer argument, R_{d1} holds the final pointer and R_{d2} holds the value. Otherwise, the argument is either an *integer* or a *list-integer* and is contained in R_{d1}, while R_{d2} is meaningless.

On unification, it should be determined whether any pointer in a linked list treats the variable as read-only. Such indication makes the referenced argument read-only as well. To support this, the Read-Only flag (R) of the condition code is utilized. The F bit of the instruction controls the setting of R. If F is set, then if during the search one of the pointers is read-only, the R bit of CC register is set, otherwise it is cleared. Later, a conditional instruction with either the RO or NRO conditions may be used.

CARMEL-1 CONTROL AND PIPELINE

CARMEL-1 pipeline is designed for single cycle execution of almost all instructions. It consists of three stages: instruction fetch, register read and ALU computation, and register write. Each machine cycle is also the clock cycle. All instructions, except Deref, BRonTag and IfTag, are completed within a single cycle; Deref takes a variable number of cycles, BRonTag takes three cycles, and IfTag takes either one or two. Four activities can take place in parallel during each cycle: the processor internal actions, one data memory access and one instruction memory access, and a detection of whether the next instruction is a Call or a Jump. Internal forwarding helps overcome pipeline dependencies. In the following we survey the implications of various instructions on the pipeline.

Load and Store

It is extremely important in CARMEL-1 to have fast Load and Store. The ability to exploit a large register file in order to minimize memory traffic is very limited in FCP, because there are less local scalar variables that can be manipulated inside registers. Most arguments are global and structured. Thus, many data accesses in CARMEL-1 are to external memory.

Usually in VLSI RISC processors, the memory cycle is a timing bottleneck which determines the length of the machine cycle. In each cycle, the next instruction is fetched, except for Load and Store which require a second cycle for data access (Katevenis 1984). In CARMEL-1, in order to allow single cycle Load and Store, we employ memory interleaving and functional separation of data and instructions into two different memory modules. At the beginning of the data-read cycle, two addresses (for instruction and data) are generated one after the other. Both instruction and data are read into the processor at the beginning of the next cycle. On a data-write cycle, the data word is written by the processor onto the data bus, in addition to the two addresses and at the same time (see Figure 1).

If the instruction which follows the Load requires the awaited argument, another useful instruction is inserted by the compiler. If the value fetched by Load is used by the second following instruction, during the cycle while it is actually written into a register, an internal forwarding mechanism provides the value for computation. Note that a ".. *Load* R_i, *Store* R_i .." sequence does not require a delay as long as the destination address computation of Store does not use the value of R_i. However, ".. *Load* R_i, *Store* R_j .." sequence where $i \neq j$ is forbidden.

Single-Cycle Unconditional Jump

An unconditional jump instruction specifies the absolute target address within the instruction itself. Four out of 32 bits are dedicated to the opcode, and the remaining 28 bits make the absolute address. The external Jump/Call Detector (see Figure 1) identifies the Jump opcode instruction shortly after it comes out of the instruction memory. In case of unconditional transfer instructions, the bits which constitute the address are latched into the B_l address buffer, overriding the "next PC" which is sent by the processor on the address bus. Consequently, the correct next instruction is fetched. Meanwhile the instruction is also transferred to the processor, and PC is updated too. Hence, no cycle is lost, and the unconditional branch instruction is executed in exactly one machine cycle.

The single cycle Jump depends on absolute addresses provided as immediate operands within the instruction. While this scheme is not completely general, it is suitable for FCP and CARMEL-1, where all target addresses are known at compile time. Moreover, it saves the need to allocate a register for the target address. However, other addressing modes (PC-relative and register-based) are also available using the Jump-Condition instruction with an always-true condition.

Single-Cycle Call

The *Call* instruction keeps the absolute address within the instruction word, similar to unconditional Jump, and exploits the same mechanism. We decided not to save the return address in a register, because register (5-bit) specification within the instruction significantly reduces the reachable address space. Instead, we have devised a five-element *circular register stack*. The return addresses are pushed onto the stack. Consequently, the Call instruction contains a full 28-bit absolute address, and the external Jump/Call detector can be employed, making Call a single cycle instruction. In addition, this solution also simplifies the job of the compiler by eliminating register allocation for return addresses.

In our implementation of FCP, the subroutine nesting depth is limited by four, except for one recursive subroutine (*unify*). The nesting depth of *unify* depends on the structural nesting of FCP data structures, but we find it is usually five or less (when manually specified in the program) or huge (when the structure is program generated). Thus, except for *unify*, nesting of all other subroutines can be supported by CARMEL-1's five register stack.

Unify and all other system subroutines are mutually exclusive; *unify* is always called from the main program, and starts with an empty stack. Consequently, overflow can only be caused by *unify*. To handle stack overflow, an extension of the stack is managed by the processor in external memory. Upon overflow, the current bottom register is pushed onto the external stack. The reverse operation takes place on underflow (on return).

DATA PATH AND REGISTER FILE

Figure 3 describes the data path of CARMEL-1. The register file has three ports, two for register read and one for register write. The 25 registers file is organized as a single window. As explained above, FCP, unlike procedural languages, does not employ user subroutines. The system subroutines almost always use global data, so the number of arguments transferred by subroutine calls is close to zero, and very few local variables are used by subroutines. Consequently, multiwindow register files (Tamir and Sequin 1983, Katevenis 1984) are unnecessary in CARMEL-1.

CARMEL-1 does not need a large number of general purpose registers for local scalar variables. As also mentioned above, scalar integer variables are not the dominating type in FCP. The number of general registers required is even smaller because return addresses are held in the circular register stack. As a result, all register allocation is predetermined, and 25 registers suffice.

Nine registers are dedicated as special FCP abstract machine (FAM) registers (see Table 2). Some other FAM registers, which are used only during limited periods, are allocated to general purpose registers when needed. The remaining FAM registers are

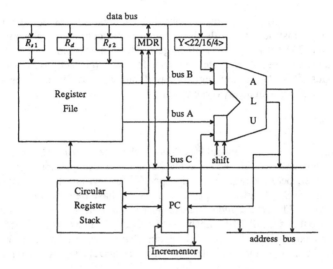

Figure 3 CARMEL-1 Data Path

used infrequently and are kept in memory.

Table 2 Dedicated CARMEL-1 Registers

HB	Heap Backtrack pointer
HP	Heap Pointer
QF	Queue Front pointer
QB	Queue Back pointer
PFL	Process Free List pointer
SFL	Suspension Free List pointer
AP	Activation (Wake-up) Pointer
STP	Suspension Table Pointer
TRP	TRail Pointer

PERFORMANCE EVALUATION

The VLSI layout of CARMEL-1 is currently being designed; actual measurements will be made once the chip is fabricated successfully. Meanwhile, an instruction level simulator has been constructed, to help us evaluate its performance, as well as investigate architectural alternatives. To utilize this investigation, the simulator computes run-time statistics such as instruction frequency and execution time distribution, memory access patterns, register utilization, type and tag distributions, and others. In addition to the simulator, a compiler, an assembler and a debugger were also written for CARMEL-1.

CARMEL-1 is designed for implementation in 1.25 μm CMOS. Thanks to the extra simple design, the critical logic path consists of only 14 gates. Using 2 nsec delay per gate as a conservative measure and leaving additional margins for phase transitions and driving internal busses and external loads, we estimate that the cycle will be 36 nsec. Since (almost) each instruction takes a single cycle, this amounts to 27 MIPS. Measuring logical inferences while executing *append*, we get 540 KLIPS with our current compiler (180 KLIPS with our previous compiler). Reported performance for some other processors running *append*, as well as some other characteristics, are shown in Table 3.

Table 3 CARMEL-1 vs. Related Processors

Processor	Lang.	Compiled/ Interpreted	R/M[1]	Perf.[2] [KLIPS]	Ref.
CARMEL-1	FCP	C	R	540 E	
PSI-II	KL0	C	M	400 M	Nakashima and Nakajima 1987
PLM	Prolog	C	M	425 E	Mulder and Tick 1987
LOW RISC	"	C	R		Mills 1987
RPM	"	C	R	526 E	Cheng *et al* 1987
OrPP	"	C	R		Biswas and Su 1985
UU	"	I	M		Woo 1985
VPI	"	I	M	30 E	Civera *et al* 1987
HPM	"	C	M	280 E	Nakazaki *et al* 1985
IPP (ECL)	"	C	M	1,000 M	Abe *et al* 1987

Note 1: Risc or Microcoded.   Note 2: Performance executing *append*.   E-estimated, M-measured.

CONCLUSIONS AND FUTURE WORK

CARMEL-1 is the direct product of an architecture-oriented analysis of FCP. We have shown how the analysis results have led us to the design of a simple and efficient processor for the language. CARMEL-1 is currently being simulated, and shows a performance of over 540 KLIPS executing *append*.

The analysis which provides the groundwork for CARMEL-1 is based on an experimental prototype software system. CARMEL-2, the successor of CARMEL-1, is currently being designed. Here we survey some of its architectural improvements over CARMEL-1.

CARMEL-2 includes ten special FCP instructions. Instead of Deref for example, an intelligent dereference instruction with tag prediction mechanism is used. Thus dereference and the following type identification are performed in parallel. In addition to Jump and Call instructions, the external detector handles the Return instruction as well, and Return is effectively executed in zero cycles. The internal circular register stack is substituted by an external one, not limited in depth. A special pipeline scheme allows "push/pop" actions to be performed "in the background," concurrently with the regular execution of instructions.

The type-identification instructions, as well as the tag separation and mapping hardware are more powerful in CARMEL-2. For example, the *cdr-coding* of lists is directly supported. *Tail-recursion* (Shapiro 1986) is supported by an internal 8-bit counter and two special instructions. A branch-based, rather then *skip* IfTag instruction was found more appropriate. More improvements were done in the VLSI level, and the estimated cycle time of CARMEL-2 is shorter.

Meanwhile, a prototype CARMEL-1 is being implemented in CMOS. Our most important goal is to investigate parallel architectures and the issues of distributed computation in FCP. We investigate shared memory and message based schemes, exploiting AND-parallelism.

ACKNOWLEDGMENTS

The invaluable help of Ehud Shapiro, the designer of FCP, is gratefully acknowledged. We also want to thank Shmuel Kliger and Avi Natan for the fruitful discussions on FCP, its compilation to CARMEL-1, and further to CARMEL-2. Avshalom Houri helped us understand the insights of his abstract machine (FAM).

REFERENCES

Abe, S., Bandoh, T., Yamaguchi, S., Kurosawa, K., and Kiriyama, K., "High Performance Integrated Prolog Processor IPP," *Proc. of the 14th Int'l Symposium on Computer Architecture*, pp. 100-107, ACM, 1987.

Biswas, P. and Su, S. H., "Design of a Processor Node to Support OR-Parallel Execution of Logic Programs," *Proc. of the Int'l Conference on Computer Design*, pp. 477-480, Oct. 1985.

Cheng, C. Y., Chen, C., and Fu, H. C., "RPM: A Fast RISC Type Prolog Machine," *CompEuro VLSI and Computers*, pp. 95-98, IEEE, Hamburg, 1987.

Civera, P., Piccinini, G., and Zamboni, M., "VLSI Architecture for Direct Prolog Language Interpretation," *CompEuro VLSI and Computers*, pp. 168-172, IEEE, Hamburg, 1987.

Gimarc, C. E. and Milutinovic, V. M., "A Survey of RISC Processors and Computers of the Mid-1980s," *IEEE Computer*, pp. 59-69, Sep. 1987.

Ginosar, R. and Harsat, A., "Profiling LOGIX: A Step Towards a Flat Concurrent Prolog Processor," *EE PUB. Technical Report Nr. 629*, Technion - Israel Institute of Technology, Haifa, Feb., 1987.

Harsat, A., "Architecture-Oriented Execution Analysis of Flat Concurrent Prolog," *Technion, Israel Institute of Technology, M.Sc Thesis*, Haifa, Israel, July, 1987.

Houri, A. and Shapiro, E., "A Sequential Abstract Machine for Flat Concurrent Prolog," *Weizmann Institute Technical Report CS-20*, Rehovot, 1986.

Katevenis, M. G. H., "Reduced Instruction Set Computer Architectures for VLSI," *ACM Doctoral Dissertation Award*, MIT Press, 1984.

Mills, J. W., "Coming to Grips with a RISC: A Report of the Progress of the LOW RISC Design Group," *Computer Architecture News*, vol. 15, pp. 53-67, SIGARCH, Mar 1987.

Mulder, H. and Tick, E., "A Performance Comparison between PLM and a M68020 Prolog Processor," *Proceedings of the 4th Int'l Conference on Logic Programming*, vol. 1, pp. 59-73, MIT press, 1987.

Nakashima, H. and Nakajima, K., "Hardware Architecture of The Sequential Inference Machine: PSI-II," *Proc. of the Int. Symposium on Logic Programming*, pp. 104-113,

IEEE, San Francisco, Aug. 31, 1987.

Nakazaki, R., Konagaya, A., Habata, S., Shimazu, H., Umemura, M., Yamamoto, M., Yokota, M., and Chikayama, T., "Design of a High-speed Prolog Machine (HPM)," *Proc. of the 12th Int'l Symposium on Computer Architecture*, pp. 191-197, 1985.

Shapiro, E., "Concurrent Prolog: a Progress report," *IEEE Computer*, pp. 44-58, August 1986.

Tamir, Y. and Sequin, C. H., "Strategies for Managing Register File in RISC," *IEEE Trans. on Computers*, vol. c-32, No. 11, pp. 977-989, Nov. 1983.

Warren, D. H. D., "An Abstract Prolog Instruction Set," *SRI Technical Note 309*, SRI International, 1983.

Woo, N. S., "A Hardware Unification Unit: Design and Analysis," *Proc. of the 12th Int'l Symposium on Computer Architecture*, pp. 198-205, 1985.

1.4 VLSI FOR PARALLEL EXECUTION OF PROLOG

Jeff Reynolds and Sergio Delgado–Rannauro

INTRODUCTION

The future exploitation of Artificial Intelligence will depend on faster implementations of languages for describing symbolic manipulation. Prolog is widely accepted as having the desired expressive power, but is often too slow for real applications.

The accepted technology for Prolog implementation involves compilation to an abstract machine, typically modelled on that of Warren (1983). For execution on existing architectures this machine may be emulated or converted into native code. Also much work has gone into designing processors which execute the sequential Warren machine directly (Dobry 1988, Nakashima and Nakajima 1987 and Fu *et al* 1987).

Though it is important to support the sequential execution of Prolog, we feel that it is even more important for the future to consider what architectural support is required for parallel execution. To hope to achieve some of the aims of AI we need to explain how large numbers of processors can cooperate together. A number of groups have been working on OR-parallelism in Prolog and published results so far (Warren *et al* 1988, Butler *et al* 1986, Hausman *et al* 1987, Reynolds *et al* 1987) indicate that the exploitation of OR-parallelism in Prolog leads to independent computations which seldom need to communicate. This is essential if a parallel architecture is not to be swamped with communications overheads. Furthermore our own experience, that of Touati and Despain (1987) and Tick (1988) indicate that logic programs exhibit good locality of data reference. However there is still the occasional need for one computation to access data structures built by another.

Most implementations of OR-parallelism to date have concentrated on shared-memory machines (Warren 1988), where this access is not too problematical. This is sufficient to exploit the current generation of multiprocessors, but sooner or later we must cope with the architectural implications of distributed data access. Accordingly in this paper we discuss an architecture for VLSI which integrates a logic processor, a page cache for remotely accessed data and a message passing unit. We also consider what hardware support could be provided for directing processors to the best available tasks.

We are going to concentrate on the processing node, and take for granted the provision of an interconnection network along the lines of the BBN butterfly (BBN 1987). The assumed capability of the network is low latency ($< 10\mu s$), uniform point to point communication.

We start with a description of aspects of the BRAVE abstract machine architecture for OR-parallel execution of logic programs and then show how the proposed physical architecture provides support for the abstract machine.

THE BRAVE ABSTRACT MACHINE ARCHITECTURE

BRAVE is an OR-parallel dialect of Prolog which has been under development at Essex University for three years. The execution mechanism of BRAVE is stack based like the Warren abstract machine, but a multiple depth-first search strategy is employed similar to the SRI model described by Warren (1987).

Prolog execution proceeds by single depth-first search of a tree, choicepoints are pushed onto a stack whenever more than one clause may match a goal. In the BRAVE execution model these choicepoints represent *potential* parallelism, but a parallel process only comes into existence if a physical processor is idle and takes an OR-choice. The crucial property of this execution mechanism is that parallelism does not expand exponentially, but rather is restricted to the number of physical processors available. Additionally, processors can work independently on their part of the search tree in a fashion which is almost as efficient as in sequential execution.

The run time data structures are organised as stacks, which branch into tree stacks whenever OR-parallel branching occurs, to permit parallel growth by separate processors. A processor builds data structures and environments at the tips of the branches of stacks in its own physical memory.

Because Prolog allows partially bound data structures a special problem occurs when such structures become shared and a processor wishes to bind a variable. The processor cannot be permitted to bind the variable directly, as this may conflict with the binding another processor wishes. The solution is to initialise all variables to a unique index as data structures are formed. Binding of a shared variable may then be performed locally in a binding stack by recording the index and the value. With this scheme writing only occurs at the tips of the tree, and as soon as an area of the tree is shared data becomes read-only. Thus we avoid the complicated locking problems of execution schemes like parallel graph reduction (Clack and Peyton Jones 1986).

Looking up the binding of a variable may require a search back through the binding stack. If the portion of the stack being searched is not local, then this access must be requested from another processor. To avoid continual search each processor maintains a software cache, the binding array, which is used for quick access to bindings for variables. The index assigned to a variable can conveniently be its actual address in this binding array, as long as all processors' binding arrays are addressed the same way. This is feasible as the binding array can be in local private memory.

Each processor also maintains a taskstack holding task descriptors or choicepoints, with references to OR-choices not yet examined. This data structure requires locking, as all processors are able to take choices from each other's taskstacks. As an optimisation only choicepoints with more than one alternative are pushed onto the top of the taskstack, so that deterministic execution does not produce choicepoints. The granularity of computation can be increased by making only an older portion of each taskstack visible to other processors, as in (Warren 1988).

TASK SCHEDULING

We consider a processor as being able to perform one task at a time for reasons which will be mentioned later. Execution proceeds with each processor fetching and executing instructions from its own copy of the program code. Whenever it encounters an instruction which indicates OR-choices it pushes a choicepoint on its own taskstack. Typically when a task fails, a processor makes a local search for a new task from the *top* choicepoints onto its own taskstack. Thus the *primary strategy* for scheduling tasks is *depth-first*. Occasionally a subtree becomes exhausted and a further strategy is required which minimises the costs both of finding a task, and switching to it. Tasks are sought from other processors by asking them to search their own taskstacks. They perform this remote search from the *bottom* of their taskstacks. This *secondary strategy* is *breadth-first* and directs processors to explore subtrees which split as close to the root of the search tree as possible. The combination of these two strategies is to promote large, independent computations with the minimum of mutual interference.

A major factor in switching tasks is the cost of reconstituting the binding array to correspond to a new position in the search tree. The binding array grows as a particular execution path is followed down towards a leaf. On failure it must be "wound back" to the last OR-choice before execution can proceed. This is because bindings made on one OR-branch may not be valid on another. If a processor needs to totally abandon a sub-tree and to move somewhere else, it may do so by winding back to the common ancestor node, and then moving down the tree, installing bindings in its binding array form the binding stack as it goes.

A strategy which helps processors switch to closely related tasks also has the important benefit that related data structures are built in the same physical memory. This improves locality of reference and furthermore part of the contents of a data cache will remain valid after a switch to a closely related task.

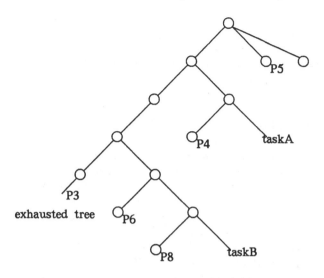

P3's active children list: [P6, P4, P5]

Figure 1 A partially–complete proof tree

To facilitate moving around the tree to look for work each choicepoint has a link to its parent. For a processor to find an optimal next task it is also useful to have information about which processors took tasks from choicepoints that it produced. This information is kept in a list referred to as the active children list, which is grown by appending processor numbers to the front, as they become active in the subtree. Processors are deleted from this list when they cease to be active. Figure 1 shows the view processor P3 has of a proof tree at a particular point in a proof where it has exhausted a subtree. It can interpret its active children list as showing what other processors are working on portions of the tree branching off its own nodes back to its initial node.

As we shall see in the next section a processor must not commence a task above any of its active children. In Figure 1 P3 must search its first child P6's active children list to find P8 which has taskB for it, rather than taking taskA from P4. Hence a processor's secondary strategy for seeking work: ask the first processor in its active children list if it has work, and if it has none ask it to go through its own active children list asking for a task. If none of the children have work then *their* children must be searched, and so on. If no work is found in all the descendants, the search must be repeated until either a child generates new work, or all children finish.

If a processor has no active children its needs a *tertiary strategy*. It goes to the parent of its initial node and starts searching further back up the tree, using its parent's active children list.   Potentially it can search the entire tree, and this could be more costly than picking a task at random from another processor, then recreating the binding array entirely.   The scheduling of the best next task is a complicated problem and is the subject of much current research (Calderwood 1988).

MEMORY MANAGEMENT

A good feature of a stack based sequential execution is that new structures grow on top of older ones, and when space recovery occurs the most recent structures are always removed first.   In the case of success or failure a processor recovers space on the stacks by using information in the choicepoints. This mechanism eliminates the need for elaborate garbage collection algorithms. However in parallel execution there is the potential problem of 'gaps' in the stacks.   This arises when a parent processor cannot recover space on a stack because a child processer still requires it.   If the parent continues to build the stack a gap is created when the child terminates and no longer requires the stack portion.   This gap can only be recovered when the parent finally fails and can recover space back to a point above.

Hermenegildo (1987) showed how this problem can be avoided in a parallel stack implementation, provided an appropriate scheduling algorithm is used. In order to preserve the required data structure precedence, a processor must not obtain work from elsewhere in the search space if it has active offspring in the subtree below it.   The active children list from the previous section provides this information.   This restriction does not impose a heavy performance penalty on the machine and there are some benefits: a clear identification of the set of processors working in any subtree assists in terminating unwanted computations.   This is required for implementing built-in predicates like *not*.

THE PHYSICAL ARCHITECTURE

Studies of distributed access using transputers (Reynolds and Lyons 1987) have shown that the overheads can be high.   Detecting an external access and then initiating a message can cost ten times a normal direct access to RAM. It is not the speed of the communications link which is the problem in this case, but the requirement for processor activity. We therefore propose an architecture which minimises the overheads to the main logic engine of detecting and performing distributed accesses.   This is achieved by providing a cache for data retrieved by remote access, and off-loading the actual burden of communication into an independent message unit.   We propose these three elements be incorporated in one chip, which is externally connected to local memory (see Figure 2).

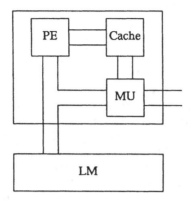

Figure 2 The physical processor

Our AI machine thus consists of a set of identical PNs, connected by a uniform interconnection network. Each PN contains a von Neumann machine with a high bandwidth between processor and memory. A global address is used to refer to data structures distributed over the machine, which can easily be interpreted as PN number + local address. The local memory of a node is large (megabytes) and logically split in two: a private section which is addressed the same way in each node, and a unique global address range. The global portion is accessible to both the processing element and the message unit.

We propose a processor which does not multi-task for a number of reasons: firstly the hit ratio of the cache is bound to be affected since the different active paths being pursued by one processor will not be closely related; secondly the context for a logic machine is quite large and multiple register sets might not be possible; thirdly, we see the technology of interconnection networks as having improved enough to give low latency.

Our architecture has something in common with the DOOM architecture being developed in ESPRIT project 415A (Odijk 1987), the proposed FLAGSHIP architecture (Watson *et al* 1987) and the BBN butterfly (BBN 1987). It differs considerably in philosophy from architectures which place processors and memory on the opposite sides of the communications network, such as the Aquarius multiprocessor (Fagin and Despain 1987). The FLAGSHIP project has moved from advocating a remote processor-memory structure, as with the ALICE prototype, to closely coupling processors and memory as we advocate. Below we look at the processor in more detail.

The processing element (PE)

The processing element is a special logic processor for the execution of the BRAVE abstract machine. There are well-defined requirements in terms of registers for the machine state:

> argument registers for parameter passing and temporary values
> stack pointers
> continuation state
> utility pointers for unification

The state can fit in the VAX 8600 (Gee *et al* 1987), but not quite in the MC68020. The requirements for managing a number of stacks fits well with existing architectures, which have auto-incrementing user stack pointers.

In common with LISP and other languages for symbolic manipulation, data typing through tagging is highly desirable, but only provided currently on certain machines. A memory which is 32 bits plus 8 bits of tag also allows for reference counting or other garbage collection schemes (Peyton Jones 1987). The MC68020 performs 50 klips as opposed to 150 klips for a machine of a similar cycle time, primarily due to its poor tag handling (Mulder and Tick 1987).

The PE is responsible for recognising local and external memory accesses prior to the request of any memory transaction.

The cache

This is a table look-aside buffer with the pages included in place. The cache associatively searches the look-aside buffer as soon as the PE signals an external memory transaction. If there is a page fault the cache requests the page from the message unit. Note that the data stored in the cache is read-only because it belongs to the shared portion of the computational tree. This avoids the problems of data coherence between caches.

A small page size of about 32 bytes is thought best: Tick (1988) has shown that memory references cluster well in areas of that size, and page fetch latency is kept low. Fetching larger pages is wasteful if a significant portion of the page is not used. We believe that demand paging will be more beneficial than prefetching data.

The message unit (MU)

The message unit is a co-processor responsible for all the message processing in the node. It must cope with any request from the PE or cache to send messages, as well as independently responding to external requests. Small FIFO buffers are used for incoming and outgoing messages.

A page request from the cache causes the MU to create a message requesting that page from the PN that owns it. On receipt of the page the MU writes it into the cache. The transaction has to be completed for the PE to continue execution.

External memory management functions

It is useful to make PNs help reclaim storage in memory areas non-local to them. The space used to hold a computation subtree can only be recovered when all the PNs working there have completed the search. In order to signal this, children who complete a task must detect when their parent is non-local and send a special memory management message to the parent. A MU receiving such messages can handle them automatically, decrementing reference counts in the choicepoints concerned. As it detects choicepoints are "dead" it generates its own memory management messages for the MUs of other PNs.

Maintenance of the taskstacks and the active children list

The message unit takes responsiblity for allowing tasks to be taken by remote processors, so that PNs seeking work communicate with the MUs of other PNs. If an MU gives a task from a choicepoint to a particular PN it also maintains the active children stack. It also is responsible for deleting children from the active children list when they report that they are no longer active.

CONCLUSION

The main reason for choosing to exploit OR-parallelism in logic languages is that it gives rise to good opportunity for independent computation, minimising communication. This is large-grained parallelism where the performance of the separate processors can be close to that of sequential implementations. However distributed access to data structures is not totally avoided, and on failure the effort of finding a new useful computation can be large. We have outlined an architecture to support these two activities. Three elements of the processing node: the PE, MU and cache could be fabricated on a single chip with current integration levels. The local memory needs to be of the order of a few megabytes for typical AI applications, thus we feel an interface for external RAM is still desirable.

We have emphasised the improvement of the processing unit, but undoubtably VLSI will contribute to improving the interconnection network as well. We look forward to seeing the next innovations in VLSI directed towards symbolic processing.

REFERENCES

BBN, *Butterfly Scheme Reference Manual.* Cambridge, MA: BBN Advanced Computers, 1987.

Butler, R., Lusk E.L., Olson R. and Overbeek R., "ANLWAM: A Parallel Implementation of the Warren Abstract Machine", *Internal Report*, Argonne National Laboratory, USA 1986.

Calderwood A., "Scheduling Or-parallelism in Aurora - the Manchester Scheduler", submitted *Int. Conf. on Fifth Generation Computers*, Tokyo, 1988.

Clack C. and Peyton Jones S.L., "The Four-Stroke Reduction Engine", in *Proc. ACM Conference on LISP and Functional Programming*, Boston, 1986.

Dobry, T.P., *A High Performance Architecture for Prolog*. Norwell, MA: Kluwer Academic, 1988.

Fu H.C., Cheng C.Y. and Chen C., "PIER: A VLSI Prolog Inference Engine on a RISC", in *Proc. IEEE Int. Symp. Circuits and Systems*, pp 938-941, 1987.

Fagin B.S. and Despain A.M., "Performance Studies of a Parallel Prolog Architecture", in *Proc. 14th Annual Symp. Computer Architecture*, pp 108-116, 1987.

Gee J., Melvin S.W. and Patt Y.N., "Advantages of Implementing Prolog by Microprogramming a Host General Purpose Computer", in *Proc. 4th Int. Conf. on Logic Programming*, pp 1-20, Melbourne, 1987.

Hausman B., Ciepielewski A. and Haridi S., "OR-parallel Prolog Made Efficient on Shared Memory Multiprocessors", *Research Report SICS 87006*, Swedish Institute of Computer Science, 1987.

Hermenegildo M.V., "Relating Goal Scheduling, Precedence, and Memory Management in AND-parallel Execution of Logic Programs", in *Proc. 4th Int. Conf. on Logic Programming*, pp 556-575, Melbourne, 1987.

Mulder H. and Tick E., "A Performance Comparison Between the PLM and an MC68020 Prolog Processor", in *Proc. 4th Int. Conf. on Logic Programming*, pp 59-73, Melbourne, 1987.

Nakashima H. and Nakajima K., "Hardware Architecture of the Sequential Inference Machine", *IEEE Symp. Logic Programming*, pp 92-102, 1987.

Odijk E.A.M., "The DOOM system and its applications: A survey of Esprit 415 subproject A, Philips Research Laboratories", in *Proc. PARLE conference*, pp 461-479, Nijmegen, Holland 1986.

Reynolds T. J., Beaumont A.J., Cheng A.S.K., Delgado-Rannauro and Spacek L.A., "BRAVE - A Parallel Logic Language for Artificial Intelligence", in *Frontiers in Computing*, pp 221-235, Amsterdam Dec. 1987.

Reynolds T. J. and Lyons D.M., "Transputers and Parallel Prolog", in *Int. Workshop on Parallel Programming of Transputer based Machines*, Grenoble 1987.

Tick, E., *Memory Performance of Prolog Architectures*. Norwell, MA: Kluwer Academic, 1988.

Touati H. and Despain A., "An Empirical Study of the Warren Abstract Machine", *IEEE Symp. on Logic Programming*, pp 114-124, 1987.

Warren D.H.D., "An Abstract Prolog Instruction Set", *SRI International Technical Note 309*, Menlo Park, CA, 1983.

Warren D.H.D., "The SRI Model for Or-parallel Execution of Prolog - Abstract Design and Implementation Issues", *IEEE Symp. Logic Programming*, pp 92-102, San Francisco, 1987.

Warren D.H.D., "The Aurora Or-Parallel Prolog System", submitted to *International Conference on Fifth Generation Computers*, Tokyo, 1988.

Watson I., Sargeant J., Watson P. and Woods V., "Flagship Computational Models and Machine Architecture", *ICL Tech. J.*, pp 555-574, May 1987.

Chapter 2

FUNCTIONAL PROGRAMMING ORIENTED ARCHITECTURES

Functional programming (FP) languages (Backus 1978) have received much attention in recent years; part of this interest is due to the advantages of these languages over the imperative ones. In this programming style, inputs are mapped onto outputs, the mapping being determined by a function; the program is therefore a function in the mathematical sense. Functional languages advantages include: *referential transparency* the output values depend only on the function's textual context, *no side effects* since the output is only a function of the inputs, and *verifiable programs*, verification can be based on function proof. However, these features can be overshadowed by their slow execution on von Neumann style architectures; therefore, it is necessary to develop new architectures. Vegdahl (1984) surveys proposed architectures for functional programming.

In this chapters two architectures are studied: the first is an architecture for functional and logic programming while the second is an architecture for knowledge bases.

AN ARCHITECTURE FOR FUNCTIONAL PROGRAMMING

An architecture for functional programming languages is described by O'Donnell §2.1. The applicative programming system architecture (Apsa) is a data–parallel fine–grain machine. A binary tree interconnection network connects a linear sequence of memory cells. This binary tree provides short data paths, extensible data structure, and low latency. Algorithms with hierarchical locality can be efficiently implemented in this architecture. Data structures that are supported by this machine include functional aggregates, a combined list/vector structure, string reduction and graph reduction. A working VLSI prototype of Apsa has been built and an emulation on the NASA massively parallel processor (MPP) has been implemented.

A FUNCTIONAL PROGRAMMING APPROACH FOR NATURAL LANGUAGE PROCESSING

Frost *et al* (§2.2) propose a systolic array architecture for the integration of natural language processing and data base manipulation. The theoretical basis of Montgue's method –which is used in this approach– has many similarities with the pure functional programming languages. The knowledge manipulation is done by functions which are based on the MIRANDA functional programming language. The answer to a query is processed by a *parser function* which creates parse trees of tagged words and phrases and an *evaluator function* which answers the query. Natural language parsing and interpretation constitute a good problem domain for this approach and architecture since the required knowledge base is large and relatively static and speed of execution is important.

References

Backus, J., "Can Programming Be Liberated from the von Neumann Style? A Functional Style and its Algebra of Programs," *Communications of the ACM,* vol. 21, pp. 613–641, August 1978.

Vegdahl, S. R., "A Survey of Proposed Architectures for the Execution of Functional Languages," *IEEE Transactions on Computers,* vol. C–23, no. 12, pp. 1050–1071, December 1984.

2.1 SUPPORTING FUNCTIONAL AND LOGIC PROGRAMMING LANGUAGES THROUGH A DATA PARALLEL VLSI ARCHITECTURE

John O'Donnell

INTRODUCTION

Since VLSI is the dominant new technology for constructing computer hardware, it is important to find effective ways to use it for a variety of applications. Several approaches for using VLSI to support artificial intelligence research are currently under study:

- Use RISC (reduced instruction set computers) to obtain a good balance between cost and performance in the design of conventional sequential architectures.

- Since a processor can fit on one chip, it is now possible to build large multiprocessor systems. Multiprocessors contain an interconnection network that supports communications among the processors, and some (but not all) of the systems implement a global shared memory. Several multiprocessors are commercially available, and they typically contain between 10 and 100 processors.

- Custom VLSI machines can be designed to execute AI algorithms directly. For example, there is current research on using VLSI hardware to implement connectionist AI architectures. This can be much more efficient than building a general-purpose multiprocessor and then programming it to simulate a connectionist network.

- The characteristics of VLSI technology can be exploited in order to build very fast processors for functional and logic programming languages. Any AI application written in such languages would benefit from this approach.

The purpose of this chapter is to describe a parallel VLSI architecture that follows the last approach. Apsa (applicative programming system architecture) is a massively parallel VLSI architecture which is designed specifically to support the implementation of declarative programming languages, including functional and logic languages. In addition, the architecture is well-suited for direct implementation of some application algorithms.

A working VLSI prototype of Apsa has been completed, and is described by O'Donnell, Bridges and Kitchel (1987). In addition, Apsa can be implemented with reasonable efficiency on general purpose SIMD machines, and O'Donnell (1987) describes a working emulation on the NASA Massively Parallel Processor (MPP). The current VLSI prototype demonstrates the architecture's feasibility, but it does not contain enough processing elements to run realistic algorithms. Although the MPP implementation is slower, it supports 16,384 cells. Further work is in progress on both of these implementations.

The remaining sections in this chapter discuss the characteristics of VLSI that are relevant to Apsa, describe the architecture, illustrate how it supports the implementation of declarative languages, and discuss the kinds of algorithm that work well on Apsa.

VLSI AND FINE GRAIN PARALLELISM

Mead and Conway (1980) discuss several characteristics that lead to efficient VLSI circuits. In particular, the following are relevant to the design of VLSI architectures for artificial intelligence:

- The architecture should be highly regular, avoiding the use of "random wiring" and "random logic".

- Wires should be as short as possible, because otherwise the wires can account for too much area in the VLSI layout.

- Parallelism is the best way to gain speed through additional circuitry.

A good way to achieve these characteristics is to design a fine-grain parallel architecture consisting of a regular network of small, simple processors. Since each processor is very small, the organization should be SIMD (i.e., a global controller issues instructions to each processor, removing the need to place a complex control unit in every processor). An interconnection network is also needed to provide for communications among the processing elements.

The earliest SIMD architectures were associative machines, sometimes called "content addressable parallel processors" (Foster 1976). Recent SIMD machines that rely on VLSI technology include the Massively Parallel Processor (Potter 1985), the Connection Machine (Hillis 1985), and the Applicative Programming System Architecture (described in this chapter).

These SIMD architectures are very similar in the capabilities of their individual processing elements. The key differences lie in their interconnection networks:

- The Massively Parallel Processor allows direct communication only between adjacent processing elements (which are arranged in a two-dimensional square grid). This organization is intended to support numerical algorithms on arrays, and it is less efficient on the irregular data structures created by functional and logic programming languages.

- The Connection Machine provides a hypercube network called the router which allows parallel communication between arbitrary pairs of processing elements. Message conflicts can occur, but the hypercube network is rich enough to allow many messages to be sent in parallel without conflict. Steele and Hillis (1986) describe a data parallel variant of Lisp for the Connection Machine.

- The Applicative Programming System Architecture uses a tree-structured interconnection network. This is an unusual choice, because there exist some algorithms that would suffer from a bottleneck near the root of the tree. However, Apsa is not intended to execute *all* parallel algorithms efficiently. Instead, it is designed to execute a key set of data structure algorithms extremely quickly, and those algorithms do not suffer from a bottleneck near the root of the tree. Magò (1979) and Magò and Middleton (1984) use a tree structured machine for a parallel implementation of FFP, although the architecture and algorithms of that system are very different from Apsa.

It seems clear that any SIMD architecture intended for artificial intelligence applications will need some long-distance data paths within the interconnection network. That means that the Massively Parallel Processor's interconnection network is not adequate for AI. The tradeoff between Apsa and the Connection Machine is very subtle: Apsa is more likely to encounter message conflicts, but the Connection Machine's interconnection network is much more expensive, has higher latency (i.e., it is slower) and does not scale up as easily to extremely large numbers of processing elements. Much more experience with both systems is needed in order to assess their relative capabilities.

DATA PARALLELISM AND PROCESS PARALLELISM

Apsa provides two distinct forms of parallelism in declarative languages, *data parallelism* and *process parallelism.* Furthermore, it is often possible to exploit both forms of parallelism at the same time.

Declarative programs spend a large portion of their time performing expensive operations on data structures. This is primarily because functional and logic languages require flexible structures like lists, continuations, environments, shared structures, user-defined abstract data types, etc. Such data structures are usually implemented in a heap, and the program must follow many pointers in order to traverse them.

It is possible to use sequential hardware to speed up heap data structure operations. One method is to design a special processor that can follow linked data structures as fast as the RAM storage allows. Conventional processors require a large amount of overhead while traversing linked structures (fetching and decoding instructions, etc.), while a special structure-traversing controller can avoid this overhead. The Grip machine (graph reduction in parallel) uses this method successfully (Peyton Jones *et al* 1987).

Apsa uses parallel hardware to perform efficient heap operations. The basic idea is to use parallelism to reduce the number of instructions that must be executed, instead of using parallelism to execute many instructions at the same time. This is done by microprogramming Apsa's intelligent memory (called the Parallel Structure Machine) to perform data structure traversals in a single unit of time.

The Apsa garbage collector contains a typical illustration of data parallelism. The garbage collector must frequently take a pointer to an object x that is known to be accessible (hence non-garbage), and mark every other object that can be reached by following pointers from x. Suppose that x is the list $(a\ b\ c\ d)$, so that the elements a, b, c and d must all be marked. A conventional garbage collector would mark the elements in the list one at a time, following a pointer from each element to the next. Apsa tries to represent linear lists in a set of adjacent memory cells (this technique is similar to *cdr-coding*). A later section shows how Apsa implements this algorithm.

Process parallelism results from breaking a problem into independent subproblems, and solving each on a separate machine. SIMD machines appear to be incapable of supporting process parallelism: Independent processes in a multiprocessor system are unlikely to execute the same instruction simultaneously, but in a SIMD machine all the processing elements must perform the same action simultaneously.

The way to obtain process parallelism with a SIMD machine is to design the instruction set to perform extremely general operations which all processes must frequently use. For example, a functional language machine based on combinator reduction might have an instruction that applies a supercombinator to a set of arguments. Even though the processes are performing different computations, they will all need to execute a supercombinator application. Therefore a supercombinator application instruction can achieve process parallelism even though it can be implemented using data structure parallelism. (Peyton Jones (1987) gives a thorough explanation of supercombinator reduction.) It may be possible to use similar techniques for logic languages.

Declarative programming languages provide ample opportunity for combining data parallelism and process parallelism. Current implementations of logic languages perform complex data structure traversals in addition to creating many processes (van Emden 1984). Functional language implementations spend much of their time working with combinator graphs, lists, and user-defined abstract data types, and they also often create many independent processes. An architecture that can support both forms of parallelism should be able to give excellent performance for declarative languages.

THE APPLICATIVE PROGRAMMING SYSTEM ARCHITECTURE

At its highest level of organization (Figure 1), Apsa consists of a conventional sequential computer called the *Control Processor* connected to an intelligent memory called the *Parallel Structure Machine*. The Control Processor executes the declarative language interpreter, while the Parallel Structure Machine supports data parallelism. The Parallel Structure Machine is the only part of Apsa that uses custom VLSI circuitry.

The Parallel Structure Machine is a SIMD architecture that contains a set of

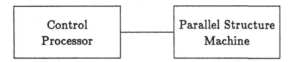

Figure 1   Organization of Apsa — Applicative Programming System Architecture

processing elements that can communicate through a tree-structered interconnection network (Figure 2). There are two kinds of processing element: *cells* are processing elements that are leaves in the tree, while *nodes* are processing elements that are not leaves. Permanent data structures are usually stored in the cells, while the nodes are used primarily for communications and logic operations that involve more than one cell.

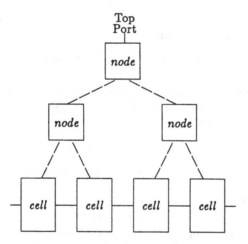

Figure 2   Organization of the Parallel Structure Machine

The Parallel Structure Machine is controlled by a sequential algorithm running in the Control Processor. When that algorithm issues an instruction to the Parallel Structure Machine, each processing element independently executes the instruction. The processing elements can perform arithmetic, logical and data movement instructions. Every processing element also contains a Mask register that prevents it from executing conditional instructions unless the Mask contains 1. This is the standard way to implement conditional operations on a SIMD architecture.

The most important communication operation within the Parallel Structure Machine is called a *sweep*. (There is another type of communication described in (O'Donnell *et al* 1979) which is called a *shift*.) A sweep causes vertical communication through the tree. There are two sweep instructions, called upsweep and dnsweep.

An *upsweep* sends data from the cells through the nodes up to the root of the tree. The execution of an upsweep consists of the following actions:

1. Each cell computes a local value and places that value on its *to* port (top output).

2. When a node receives input values on its *li* port (left input, from the left subtree) and its *ri* port (right input, from the right subtree), the node uses a function *f* to compute a value which it places on its *to* port (top output):

$$to \quad = \quad f(li, ri).$$

The function *f* is supplied by the Control Processor as an operand to the upsweep.

3. While the node is computing its *to* value, it also latches the *li* and *ri* inputs in two local registers. This does not affect the upsweep, but the saved values are frequently used during a subsequent dnsweep.

4. The Main Controller reads the *to* value produced by the root node.

A *dnsweep* instruction sends data from the root, down through the nodes, to the cells. Its execution consists of the following steps:

1. The Main Controller produces a value which becomes the top input *ti* to the root node.

2. As each node receives its top input value *ti*, the node uses a pair of functions f_l and f_r supplied by the Main Controller to compute the left output value *lo* and the right output value *ro*:

$$lo \quad = \quad f_l(ti, li, ri),$$
$$ro \quad = \quad f_r(ti, li, ri).$$

The *li* and *ri* values from the previous upsweep are available to these functions because the node saved them during the upsweep (in step 3).

3. When a cell receives its top input value *ti*, it stores the value locally.

An important property of the Parallel Structure Machine is that *the upsweep and dnsweep microinstructions execute in one hardware clock cycle.* The sequences of operations described above take place purely within combinational logic circuitry; they do not require any sequencing. In contrast, most tree multiprocessors require $\log n$ clock cycles to send information from the root to the leaves, where there are n leaves. The Parallel Structure Machine does this in 1 clock cycle.

DATA STRUCTURES FOR DECLARATIVE LANGUAGES

The instruction set of the Parallel Structure Machine supports efficient implementation of a wide variety of data structures. Examples include a combined list/vector structure, functional aggregates, continuation stacks, string reduction and graph reduction. Since many declarative language implementations spend much of their time manipulating complex data structures, this approach can greatly improve overall system performance.

Data structure representations for conventional machines are often either completely fixed (for example, arrays) or else distributed arbitrarily through the memory (for example, heaps). Data structures for the Parallel Structure Machine usually fall between those extremes. Since pointers are allowed, and it is also possible to insert new elements into any structure, these data structures have much of the flexibility of heap representation. However, the algorithms gain much more parallelism when the representation exploits locality. Therefore typical data structures contain a hierarchy of compact data structures.

In order to obtain as much parallelism as possible, it is important for the data structure representation to preserve locality. Therefore the basic kind of data structure is a "compact linear structure", a set of words stored in consecutive memory cells. Compact linear structures can be used to represent arbitrary lists; invisible pointers are needed to link shared structures together.

A *compact linear structure* is a sequence of words that are stored in a consecutive sequence of memory cells. Each cell holds two special flags, *first* and *last*. The cell that contains the first word of a linear structure must have its *first* flag set, while the cell that contains the last word has its *last* flag set.

The rest of this section illustrates programming technique for the Parallel Structure Machine with an example taken from a garbage collector, a crucial component of most declarative language implementations. The algorithm marks a number of data structure elements at the same time, while garbage collectors running on sequential computers can mark only one element at a time.

Suppose that the garbage collector has marked the first word in a linear structure (such as a list), and next it needs to mark all the remaining words in the structure. Conventional machines would do this by iterating down the list, requiring time proportional to the size of the structure. However, the Parallel Structure Machine can mark all the words in the structure in constant time; the algorithm requires a single upsweep followed by a single dnsweep.

Figure 3 shows a snapshot of part of the memory during execution of the garbage collector. There are three compact linear structures, each one beginning with a *first* element and ending with a *last* one. One of the elements has already been marked by the garbage collector, indicated with a "▬▬" symbol.

Figure 3 Initial representation — one marked list

The collector must now mark all the remaining elements in the structure whose first element is marked. It does this by executing the *mark-structure* operation. This illustrates parallelism within a single data structure, since many structure elements have been marked simultaneously. Figure 4 shows the result.

Figure 4 Result of executing mark-structure

The garbage collector also exhibits parallelism among independent data structures. For example, suppose that we begin with the same initial representation as before, except that the first element of all three of the lists has been marked (Figure 5).

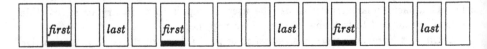

Figure 5 Initial representation — three marked lists

When the collector executes the mark-structure operation, all the elements of all three of the compact linear structures will be marked (Figure 6).

Figure 6 Result of executing mark-structure

The representation we are using guarantees that every linear structure begins with a cell marked *first* and ends with a cell marked *last*. The basic idea of the implementation of mark-structure is that a node can detect three special kinds of substructure:

- A *head* is a sequence of cells where the first cell has *first* set, and none of the other cells has either *first* or *last* set.
- A *middle* is a sequence of cells which have neither *first* nor *last* set.
- A *tail* is a sequence of cells where the last cell has *last* set, and none of the other cells has either *first* or *last* set.

A linear structure can be partitioned into a head followed immediately by a tail. Furthermore, a head followed by a middle forms a larger head, while a middle followed immediately by a tail forms a larger tail.

The Parallel Structure Machine implements the mark-structure operation with an upsweep followed by a dnsweep. During the upsweep the nodes decide whether their subtrees contain a head, a tail or a middle. During the dnsweep the nodes generate flags that determine whether each cell is a member of a marked structure. If a node sends $mark\_h=1$ to a subtree, that means the subtree contains a head that must be marked. Similarly, $mark\_t$ controls the marking of tails.

First, each cell uses the following equations to determine whether the value it contains is a head, middle or tail. The cell then sends these values up the tree, beginning the upsweep. If the cell is both the first and last in a linear structure, then the structure is a *unit* which does not require any communication through the tree.

$$
\begin{aligned}
head &= first \times \overline{last} \\
middle &= \overline{first} \times \overline{last} \\
tail &= \overline{first} \times last \\
unit &= first \times last
\end{aligned}
$$

During the upsweep, each node receives *li* inputs from its left subtree and *ri* inputs from its right subtree. The node uses these values to compute the *to* top outputs which it sends up the tree. The node also computes group, a local value which it will use later during the dnsweep.

$$
\begin{aligned}
head &= ri\_head + li\_head \times ri\_middle \\
middle &= li\_middle \times ri\_middle \\
tail &= li\_tail + ri\_tail \times li\_middle \\
group &= li\_head \times ri\_tail
\end{aligned}
$$

After the upsweep has completed, each node knows what is below it. That information must be combined with knowledge about what is to the left or right of it, and the dnsweep provides that information. To start the dnsweep, the main controller provides the top inputs to the root node. These are both zero because there are no heads or tails to the left or right of the entire memory structure.

$$
\begin{aligned}
ti\_mark\_h &= 0 \\
ti\_mark\_t &= 0
\end{aligned}
$$

During the dnsweep each node uses the following equations to compute the markh and markt messages which it must send to the left and right subtrees.

$$
\begin{aligned}
lo\_mark\_h &= group + ti\_mark\_h \times ri\_middle \\
lo\_mark\_t &= ti\_mark\_t \\
ro\_mark\_h &= ti\_mark\_h \\
ro\_mark\_t &= group + ti\_mark\_t \times li\_middle
\end{aligned}
$$

When the cells receive their top inputs, they set their local mark flag:

$$
mark \Leftarrow mark + unit + ti\_mark\_h + ti\_mark\_t
$$

Figure 7 shows in detail the values computed during a mark-structure. Each node is represented by a box with three regions. The top region shows the top outputs (*to*) computed during the upsweep. The head, middle and tail flags are abbreviated h, m and t. Below that are two regions that show the left output (*lo*) and right output (*ro*), which are computed during the dnsweep. Here the mark_h and mark_t flags are abbreviated as h and t.

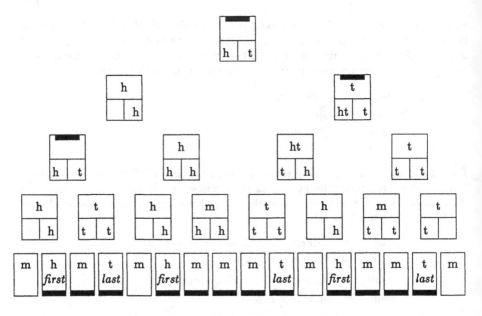

Figure 7 Execution of mark-structure

CONCLUSION

Apsa is a massively parallel VLSI architecture for implementing declarative programming languages. The key part of Apsa is a Parallel Structure Machine, which uses a large set of simple processing elements connected with a combinational binary tree network to implement key data structure operations in parallel. The algorithms for Apsa exhibit both "data structure parallelism" and "process parallelism".

The Parallel Structure Machine is well suited for VLSI implementation because its architecture is extremely regular, it contains a relatively small number of long wires, it is easily extensible to large numbers of processing elements, and a very powerful machine can be constructed by replicating small subsystems. In contrast, it is much harder to make a conventional processor more powerful simply by making it bigger.

There are two complementary paths in applying VLSI to AI. The first is to build special purpose architectures to execute specific algorithms, and the second is to build special purpose architectures to execute declarative programming languages. The first path can lead to better performance, but is much less flexible. It is very undesirable to be forced to modify hardware in order to experiment with an algorithm.

A middle ground between those paths is to design a parallel architecture that can be programmed to support a wide variety of data structure operations. This would allow the machine to support several declarative programming languages and to support a variety of specific AI algorithms. Apsa follows this middle ground by making the Parallel Structure Machine programmable.

A major problem which requires further research is to assess the tradeoffs among different kinds of interconnection networks for AI machines. It is clear that some long data paths will be needed, but we do not know yet how many. Apsa uses a combinational binary tree network which is inexpensive, contains relatively few long data paths, is easily extensible, and has very low latency (i.e., message transmission is extremely fast). In contrast, the Connection Machine contains a hypercube which has many more long data paths, is more expensive to extend to large numbers of processing elements, and has longer latency (so individual message transmission is relatively slow). Therefore the key question is: Can Apsa avoid message conflicts often enough to give better a better cost/performance ratio than the Connection Machine?

Early results with Apsa algorithms show that the binary tree network is surprisingly effective. The reason is that a binary tree machine can transmit many messages in parallel, as long as the intervals between sender and receiver do not overlap. In practice, many parallel algorithms have a natural hierarchical structure, and it is possible to exploit that structure by representing senders and receivers close to each other in the set of memory cells.

The Connection Machine should perform better on algorithms that do not posses any natural structure. For example, if a neural network has connections between randomly chosen neurons, then the Connection Machine should be a more efficient simulation host.

Further research in declarative language implementation and artificial intelligence algorithms will be needed to answer these detailed questions about architecture. Nevertheless, it is already clear that massively parallel machines with extremely fine grain provide an extremely effective way to use VLSI technology to support AI.

References

Foster, C. C, *Content Addressable Parallel Processors.* New York: Van Nostrand Reinhold Co., 1976.

Hillis, W. D., *The Connection Machine.* Cambridge, MA: The MIT Press, 1985.

Magò, G., "A Network of Microprocessors to Execute Reduction Languages," *Int. Journal of Computer and Information Sciences,* vol. 8, pp. No. 5 and No. 6, 1979.

Magò, G. and Middleton, D., "The FFP Machine — a Progress Report," in *Proc. International Workshop on High-Level Language Computer Architecture,* pp. 5.13–5.25, 1984.

Mead, C. and Conway, L., *Introduction to VLSI Systems.* Reading, MA: Addison-Wesley, 1980.

O'Donnell, J. T., "Parallel VLSI Architecture Emulation and the Organization of APSA/MPP," in *Proc. First Symposium on the Frontiers of Massively Parallel Scientific Computation*, pp. 75–84, 1987.

O'Donnell, J. T., Bridges, T. and Kitchel, S. W., "A VLSI Implementation of an Architecture for Applicative Programming," in *Proc. Conference on Frontiers in Computing*, pp. 315–330, 1987.

Peyton Jones, S. L., *The Implementation of Functional Programming Languages*. Englewood Cliffs, NJ: Prentice-Hall, 1987.

Peyton Jones, S. L., Clack, C., Salkild, J. and Hardie, M., "GRIP — A High-Performance Architecture for Parallel Graph Reduction," in *Proc. Functional Programming Languages and Computer Architecture*, pp. 98–112, 1987.

Potter, J. L, *The Massively Parallel Processor*. Cambridge, MA: The MIT Press, 1985.

Steele, G. L. and Hillis, W. D., "Connection Machine LISP: Fine-Grained Parallel Symbolic Processing," in *Proc. 1986 ACM Conference on Lisp and Functional Programming*, pp. 279–297, 1986.

van Emden, M. H., "An Interpreting Algorithm for Prolog Programs," in *Implementations of Prolog*, J. A. Campbell (ed), Chichester: Ellis Horwood, pp. 93–110, 1984.

2.2 TRANSLATING DECLARATIVELY SPECIFIED KNOWLEDGE AND USAGE REQUIREMENTS INTO A RECONFIGURABLE MACHINE

Richard Frost, Subir Bandyopadhyay and Dimitris Phoukas

INTRODUCTION

This paper is a position paper related to a project at the University of Windsor, Canada, that is concerned with the integration of natural language processing and data base manipulation. The approach that we are investigating is based on Richard Montague's ideas on natural language interpretation (Montague 1974). Two years ago, we recognized a similarity between the theoretical basis underlying Montague's ideas and that underlying many of the new "pure" functional programming languages, such as Miranda™ (Turner 1985). We have implemented a system called DMSGII in Miranda capable of answering queries, expressed in English, with respect to a first order database containing facts about the solar system. We believe that we have demonstrated that our approach facilitates the integration of natural language processors, database operations and data, and that the approach provides a useful framework for gaining new insights into the use of semantics in parsing, and the integration of extensional and intensional representations of knowledge.

In our approach, no distinctions are made between parsers, evaluators, database operators, inference rules and data. All are regarded as functions in a function space constructed over a set of entities, two boolean values and a set of strings. These functions are defined declaratively in the executable specification language Miranda. The approach has many advantages but suffers from a serious problem of inefficiency. One method of overcoming this problem, that would provide a general solution, would be to transform the executable specification to an equivalent program based on relational calculus. However, a set of rewrite rules for doing this is not yet available, and, of more importance, it is possible that in adopting such an approach, we might restrict our research horizons. In particular, the approach that we have developed may be regarded as a higher order calculus of characteristic functions of relations, and as such suggests a particular type of hardware implementation. We believe that it is worthwhile investigating a hardware solution further.

The position that we are at, therefore, is one at which we have identified a serious problem of efficiency and are considering two methods of overcoming it. One involves program transformation and the other involves special purpose hardware. We have not yet pursued either of these alternatives to any great extent, and one purpose of this paper is to prompt dialogues with both the functional programming and the advanced hardware communities in the hope that a solution to our problem can be found.

BUILDING NATURAL LANGUAGE INTERFACES TO DATABASES

Cercone and McCalla (1986) have identified two problems concerning the building of natural language interfaces to databases:

(a) "The stratified approach of doing syntactic analysis, then semantic interpretation, then query evaluation, is ineffective; techniques must be evolved to integrate syntax, semantics and pragmatics, so that whatever action is appropriate at a given time can be done".

(b) "The separation of the linguistic component from the database component sets up an arbitrary barrier which may have become counterproductive; a means of re-integrating data and language must be found".

As an example of a problem that results from a conventional approach in which linguistic and database components are separated, consider the following phrases :

"jim and john", "a man and a woman", "love mary and like john".

From a linguistic perspective, the word "and" has a similar effect in each of these phrases, it is being used to connect grammatically coordinate words or phrases. However, if these phrases were parts of queries that were translated, according to conventional approaches, into relational algebraic expressions, the word "and" would be translated into operators of quite different types. Another example concerns the word "every". Consider the following queries :

"every man thinks?", "every company employs a woman ?".

If these queries were translated into relational algebraic expressions that were evaluated with respect to a database, the evaluation process would comprise computations involving tests for set inclusion. However, given that it is known that "every man thinks", it should be possible to encode this knowledge as part of the 'intensional' definition of the word "every". The answer to the first query would then not require a possibly lengthy computation involving the sets "man" and "thinks". The conventional approach of translating the word "every" to a relational algebraic operator would appear to complicate the use of such knowledge.

OUR APPROACH

In our approach the answer to a query is obtained as follows :

(a) A parser function is applied to the query. The result returned is one or more parse trees in which all words and phrase are 'tagged' with a 'constructor' that indicates the syntactic category. For example, the word "man" is tagged, in most cases, with the constructor 'Commonnoun'.

(b) An evaluator function is applied to each of the parse trees. The results are the answers to the query. Evaluation involves translating each (syntactic category/word) pair to a single function. These functions are then applied to each other in an order given by rules, embedded in the evaluators, that correspond to the syntactic constructs in which the words appear.

As an example, consider the query "john loves mary?"

"john" translates to 'mntgran_john'.

"mary" translates to 'mntgtran_mary'.

"loves" translates to 'mntgtran_loves'.

The prefix 'mntgtran' indicates that the function is related to the "Montague" translation of the word. The answer to the query is then obtained by evaluating the following expression, where brackets indicate order of function application :

mntgtran_john (mntgtran_loves mntgtran_mary)

We shall show later that the function 'mntgtran_loves' is defined in terms of (but is not equivalent to) a function that may be regarded as the characteristic function of a binary-relation containing the data about who loves whom.

The parser function is defined in terms of more basic parsers, some of which are primitive and some of which are themselves defined in terms of other parsers. The 'glue' that is used to stick parsers together includes two higher order functions 'orelse' and 'then'. For example:

parseverbphrase = (action Intransverbphrase . parseverbphrase)
$orelse
(action Transverbphrase . parsetransverbphrase)

The dot signifies function composition and the function 'action' places the constructor that follows it (signified by a commencing capital letter) in the appropriate place in the parse tree. The dollar sign indicates that the function 'orelse' is to be used in infix mode. Another example is:

parsedetphrase = action Detphrase . (parsedeterminer $then parsenounphrase)

For each parser, there is a corresponding evaluator, for example:

evaldetphrase (Detphrase x y) = (evaldeterminer x) (evalnounphrase y)

Thus, the evaluation of a determiner phrase consisting of a determiner x and a noun phrase y, involves the application of the evaluation of x to the evaluation of y.

We have not yet given any definitions of the functions that are the translations of the syntactic category/word pairs. We shall present examples after giving a brief description of some of Montague's ideas. The important point to note is that in our approach, every syntactic category/word pair is translated to a functional expression denoting a single semantic object, and that there is no Chomsky type transformation at all. The meaning of a complex expression is computed from the meanings of its parts. It is this Fregean property of compositionality that gives our approach its power. A characteristic of the approach is that the functions into which the syntactic category/word pairs are translated have to be defined such that they fit together as do pieces of a 'mechano set' when combined in various ways. This results in some rather obscure definitions as will be seen later. The justification for this is that it enables words to be translated consistently in many of the contexts in which they appear. For example, the word "love" is translated to the same function in all of the following:

"whom does john love?", "does john love mary?",

"does every man love a woman?", "is mary love(d) by john?".

MONTAGUE'S APPROACH TO THE INTERPRETATION OF NATURAL LANGUAGE

Montague believed that any natural language, at a given point in time, is a formal language whose syntax and semantics can be defined concisely. His method for assigning an interpretation to a phrase of natural language involves four components :

(a) The natural language phrase is analyzed using a set of syntactic rules. Each syntactic rule contains information specifying how complex expressions of given syntactic categories can be constructed from simpler components of given syntactic categories. Note that more than one rule may be used to analyze a phrase. This analysis is called parsing. The output from the analysis is one or more parse trees, where a parse tree is a tree in which the expression and its components are all 'tagged' with their syntactic categories. Note that a single natural language expression can result in more than one parse tree.

(b) Each syntactic rule has a translation rule associated with it. These translation rules specify the translation of the outputs from the syntactic rules in terms of the inputs to the syntactic rules. The translation rules specify how to translate the syntax trees to expressions in an unambiguous formal language of an intensional logic called IL, i.e., a language that has a well defined model theoretic semantics. The translation of a single natural language expression into possibly many expressions in the unambiguous language is often referred to as "disambiguation".

(c) The semantics of IL tells us that each component of the expression that is output from the translation rules denotes a function in a function space constructed over a set of a constant functions corresponding to :

- Two objects, 'True' and 'False' of type boolean.
- A set of objects, E1, E2, etc. of type entity.
- A set of 'times', t1, t2, etc.
- A set of 'possible worlds', w1, w2, etc.

(d) The meaning of the IL expression is obtained by reference to the model theoretic semantics underlying IL.

Montague indicated quite clearly that translation to IL as an intermediate language was dispensable and served only to explicate the process. In our approach, IL is not used at all. We translate syntactic category/word pairs to functions whose definitions are expressed in Miranda. This is a more primitive language than IL in the sense that it does not have, for example, a built-in universal quantifier 'operator' as IL does. The semantics of Miranda are used to provide the interpretation of the English expression. Also, we have used only a few of Montague's ideas in our project. In particular, we have not included any modal or intensional concepts. This aspect of Montague's work is regarded by many as being the most significant of his contributions. We are studying how to extend our approach to accommodate such features. However, this paper is not concerned with these issues and we shall ignore them in the following.

EXAMPLES OF FUNCTION DEFINITIONS

Examples of words that DMSGII can interpret are *"planet"* and *"moon"*. These words are translated to the functions 'mntgtrans_planet' and 'mntgtrans_moon', defined as follows, where 'member s x' $\equiv$ 'x $\in$ s '.

  mntgtrans_planet x = member [E13, E19,...] x

  mntgtrans_moon x = member [E3, E8,...] x

The proper name *"mars"*, translates to

  mntgtrans_mars p = p E19.

In other words, "mntgtrans_mars" denotes a function that takes a 'property' such as that denoted by 'mntgtrans_planet' as argument and returns the value True if that property is true of the entity E19, and false otherwise. According to Montague, proper names do not denote entities. They denote functions that pick out those properties that are true of some particular entity to which the name is associated. When we hear the name "mars", for example, we do not 'picture' the entity associated with this name, but rather the intersection of all the properties that we know to be true of that entity.

Montague's interpretation of transitive verbs is one of the most difficult but ingenious aspects of his approach. Transitive verbs do not denote binary relations. They denote functions that are defined in terms of characteristic functions of binary relations defined in the set of entities. For example, the word *"orbit"* is translated to the function 'mntgtrans_orbit' that is defined in terms of a function 'orbit_rel' that is the characteristic function of the binary relation linking entities to those entities that they orbit :

  orbit_rel x y = member [(E3, E27), (E8, E19), ...] (x, y)

  mntgtrans_orbit t = g where g x = t k where k y = orbit_rel x y

We give an example later of the use of this function. Some other examples are :

English word	Function definition
is	mntgtrans_is t = g where
	g x = t k where k y = equal x y
every	mntgtrans_every p q = uniquant f where
	f x = implies (p x) (q x)
a	mntgtrans_a p q = exiquant f where f x = (p x) & (q x)
two	mntgtrans_two p q = twoquant f where
	f x = (p x) & (q x)

We assume that the following definitions are available :

  exiquant h = at_least_one_true_in (map h entityset)

  uniquant h = all_true_in (map h entityset)

  onequant h = exactly_one_true_in (map h entityset)

  implies False x = False

  implies True x = x

The function 'uniquant' takes a function such as 'mntgtrans_planet' and repeatedly applies it (this is what 'map' does) to all entities in the entityset. The resulting list of boolean values is then given as argument to 'all_true_in' which returns the value True if all values are True, and False otherwise. Functions 'exiquant' and 'onequant' are defined similarly. The functions 'implies' and '&' are the functional equivalents of the related logical connectives. The function 'mntgtrans_every' takes two functional expressions such as 'mntgtrans_moon' and 'mntgtrans_spins' as arguments, creates a function f from them and then gives this as argument to uniquant. The computation therefore involves applying f to each entity in the entityset and testing to see if all values returned are True. In this example, f will return a value of True for an entity e if 'implies (mntgtrans_moon e) (mntgtrans_spins e)' evaluates to True.

INTERPRETATION OF AN EXAMPLE QUERY

Consider the query "one moon orbits mars?". DMSGII would translate this query to the following functional expression :

mntgtrans_one mntgtrans_moon (mntgtrans_orbits mntgtrans_mars)

Evaluation of '(mntgtrans_orbits mntgtrans_mars)', according to the definition of 'mntgtrans_orbits', returns a function g where :

g x = mntgtrans_mars k where k y = orbit_rel x y

According to the definition of 'mntgtrans_mars', this reduces to :

g x = k E19 where k y = orbit_rel x y

This reduces to :

g x = orbit_rel x E19

Hence, (mntgtrans_orbits mntgtrans_mars) evaluates to a function 'g' that is of the same type as 'mntgtrans_moon'. Evaluation of the whole query, therefore, involves applying the functions 'mntgtrans_moon' and 'g' to each entity in the entityset and returning the value True for the entity if both functions return True. The list of boolean values returned by this process is then checked to see if it contains exactly one value True. If so, the answer to the query is True, otherwise it is False.

RECONFIGURABLE SYSTOLIC ARRAYS FOR QUERY PROCESSING

Queries are translated into functional expressions. The evaluation of the functional expressions comprises of two stages. The first stage involves reducing the expression to an expression containing only terms denoting entities, characteristic functions of relations, functions corresponding to boolean operators, and 'aggregate' functions described later. The second stage involves the evaluation of this reduced expression. We propose that the second stage be carried out by special purpose hardware.

The basic approach involves the determination, at run time, of an appropriate configuration of systolic arrays (Kung, H.T. 1982) and other 'logic processors' which carry out table-lookup (for evaluation of characteristic functions), and other logical operations, at high speed and in parallel. Configuration of the system involves, amongst other things, the selection of systolic arrays and the loading of these arrays with tuples from relations stored in memory. For example, the systolic array A1 corresponding to the characteristic function 'mntgtrans_planet' is loaded with entities corresponding to planets (e.g., E13, E19,....,) as shown in Figure 1.

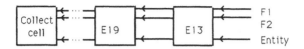

Figure 1 Array A1 corresponding to the characteristic function 'mntgtrans_planet'

Evaluation of a query is then carried out by feeding successive members of entities from the entity set {E1, E2, ... , E13, ... , E19, ... } into the input of the array. Such a sequence of entities E1, E2, ... will be called a "stream of entities". Each entity in a stream has two Boolean flags F1 and F2 associated with it. Initially, F1 is True and F2 is False, for all entities. Arrays such as A1 operate as follows: with each clock pulse, an entity from the input stream is fed into the first cell. The entities already in the array move one cell, to make room for this input. In other words, the entities E1, E2, ... propagate through the cells from right to left. When entity E13 is an input to the first cell in A1, the cell will detect a match between the input and its contents and will set F2 of this (input) entity to True. Similarly, when E19 is applied to the second cell, its associated F2 flag will be set to True. The last cell in each array is called a "collector cell" whose purpose is to set F1 to F1∧F2 and to set F2 to False. The output from the array is the stream of the entities that appeared at the input with the flag F1 updated as discussed above.

The logic processors (LPs) are simple processors that accept inputs from one or more systolic arrays or other LPs. Consider the query "which planets spin?". To process this query, the systolic array A1 as given in Figure 1 identifies all entities in the stream that are planets by setting their respective F1 flags to True. A similar array identifies all entities that spin. An LP then computes the logical AND of the respective entities produced by the two arrays. The output from this LP is a stream of entities in which the F1 flag is set to True for only those entities that are planets and that spin (see Figure 2). The processing of "or" and "not" is similar. Logical implication of the form X -> Y is another important operation. Consider the query "Every planet spins?". The configuration for the query is the same as in Figure 2. The LP now sets the flag F1 to True if the entity is not a planet or if the entity is a planet and it spins. The entities are then processed by another logic processor LP2 which, as discussed later, checks whether all incoming entities have flag F1 True. The use of a number of LPs to evaluate a complex condition is straightforward. Note that the arrays A1 and A2 in Figure 2 must have identical lengths in order that LP1 receives the same entity from A1 and A2 at the same time. Where appropriate, output from arrays or LPs may be fed into 'concentrators' that discard entities that have F1 set to False.

Logic processors may also be used to implement 'aggregate' operations such as 'all_true_in', 'exactly m true_in', 'at_least m true_in' and 'count_number_of_trues_in'. In simple conditions, such LPs do not produce an output until the last entity in the stream has been processed. More complex conditions, involving "GROUP BY" operations will be discussed later with an example. Each LP has an internal memory that may be used in aggregate operations. As an example of the use of an aggregate operation, consider, again, the query "every planet spins?". As discussed earlier, using the conFiguration shown in Figure 2, LP1 computes logical implication and the output from LP1 is fed into a processor LP2 (not shown in the figure) that produces the value True if all entities in the stream have F1=True.

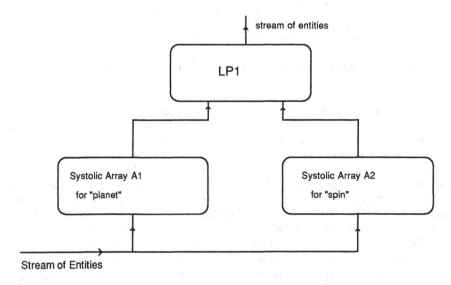

Figure 2 A configuration suitable for evaluation of queries of the form "which planets spin?"

Evaluation of queries that involve characteristic functions of binary relations is more complex in that it requires concentrators to be used, cartesian products to be generated, and may require 'group_by' aggregate operations to be performed. Consider the query "two moons orbit a planet?". The output from the first stage of query reduction, for one of the interpretations of this query, produces an expression that, after appropriate optimizations, could result in the configuration given in Figure 3. The array CP, which is described later, takes two streams S1 and S2 of entities as input and produces a stream of pairs S3=S1xS2 as result. S1 contains only those entities that are moons and S2 only those that are planets. Each pair in S3 has two flags F1 and F2 associated with it. CP is such that the pairs in S3 are "grouped" in the sense that all pairs with the same entity in the second position are produced consecutively. In this example, the system is configured so that all pairs for each planet are produced consecutively. S3 is fed into the array A4 that sets F1 to True for a pair if it is in the relation 'orbit_rel'. The stream of pairs that is output from A4 is then fed into the processor LP3. The command to LP3 is a 'group_by' aggregate command whose function may be described as follows: each group of pairs with the same entity in the second position results in one boolean output result. The output result is True if exactly two of the pairs in the group have flags set to True, and False otherwise. The stream of boolean values produced by LP3 is fed into LP4 that returns the value True if at least one of the values in the stream is True, and False otherwise.

Clearly, there are a number of obvious optimizations that could be made in this example. For instance, the arrays A1 and A3 together with their respective concentrators could be replaced by generators directly producing streams corresponding to the relations 'planet_rel' and 'moon_rel'. We have ignored such optimizations in presenting these expository examples.

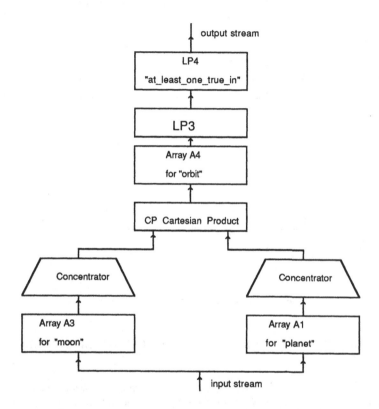

Figure 3 Evaluation of queries involving binary relations

A detailed description of the design of a systolic array to generate the cartesian product of two streams of entities is tedious. We shall discuss, with an example, how one such array works. Let a stream A consist of 3 entities a1, a2, and a3 and a stream B consist of b1, b2, b3, and b4. The leftmost column of an array with structure given in Figure 4 is initially loaded with a1, a2 and a3. With each subsequent clock pulse, a new entity from B is fed into the top cell of the leftmost column. Previous entities input from B are propagated to cells below. Whenever an entity b_i from B enters a cell containing an entity a_j from A, a pair (b_i a_j) is generated. Figure 4 shows the contents of the array after all entities from B have propagated through all cells in the left-most column. The final phase of the process involves shifting the pairs, starting with the pairs in the last row, out of the array.

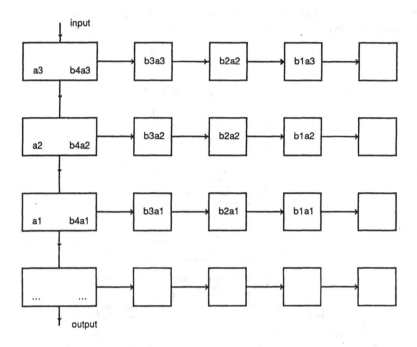

Figure 4 An Array for Cartesian Product

PROBLEMS WITH AN ALTERNATIVE APPROACH BASED ON RELATIONAL CALCULUS

Some readers may wonder why we have not based the whole of our approach on functions more closely related to relational calculus. We gave some indication in the introduction. In this section, we discuss two problems that arise in one such approach. Suppose that the 'base' functions into which syntactic category/word pairs are ultimately translated were defined as follows:

man = [E1, E2]
woman = [E3, E4]
john p = member p E1
jim p = member p E2
mary p = member p E3
susan p = member p E4
loves_rel = [(E1, [E3, E4]), (E2, [E4])]
loves m = [x | (x, y) <- loves_rel; m y]   (; denotes AND)
emptyset [] = True
emptyset s = False
is_sub_set_of s t = emptyset [x| x<-s; not (member t x)]
every p q = p $is_sub_set_of q
non_empty_intersect s t = not (emptyset [x| x<-s; member t x])

```
a p q = non_empty_intersect p q
term_and p q = f where f x = (p x) & (q x)
verbphrase_and p q = p ++ q
is_rel = [(E1, [E1]), (E2, [E2]), (E3, [E3]), (E4, [E4])]
is m = [x| (x, y) <- is_rel; m y]
who x = x
```

evaluation of the expressions into which certain queries are translated would be efficient. For example, consider the following in which we have used brackets to indicate order of function application:

Q1: john (loves mary) $\Rightarrow$ True

Q2: every man (loves (a woman)) $\Rightarrow$ True

Q3: who (loves (a woman)) $\Rightarrow$ [E1, E2]

Q4: john (loves (a woman)) $\Rightarrow$ True

Evaluation of Q1 would proceed as follows:

```
loves mary ⇒ [x | (x, y) <- loves_rel ; member y E3]
           ⇒ [x | (x, y) <- loves_rel ; member y E3]
           ⇒ [E1]
john [E1]  ⇒ True
```

A problem arises with queries such as "john is a man ?":

Q5: john (is (a man)) $\Rightarrow$ john [x| (x, y) <- is_rel ; (a man) y]

In order to maintain a consistent approach (in which queries such as Q4 and Q5 are evaluated in the same way), we cannot avoid introducing the 'is_rel' relation. What we gain in efficiency in dealing with "every" and "a" is lost in dealing with "is". The second problem is that words such as "and" have to be translated to different functions according to context, thereby losing some of the attraction of Montague's approach.

CONCLUSION

The approach to integrating natural language processing and database manipulation, that we are proposing is analogous to Hilbert's approach to number theory in which all number theoretic constructs may be built from a set of base functions using a set of construction strategies that include function composition and primitive recursion. The appeal of Hilbert's system is due to the limited number of semantic primitives that it employs. We believe that our approach has a similar appeal. The consideration that we are giving to the hardware solution to the problem is motivated, not so much by the expectation that this will provide a solution to the problem, but rather by the belief that such investigations may lead to a revision of our approach in which the problem of efficiency is not an issue.

ACKNOWLEDGEMENTS

The work reported here is partially supported by operating grants awarded by the Natural Sciences and Engineering Council of Canada to R. A. Frost and to S. Bandyopadhyay.

REFERENCES

Cercone,N. and McCalla, G. *"Accessing Knowledge through Natural Language"* . In "Advances in Computers", M. Yovits (Editor), Academic Press, NY 1986.

Kung, H.T. *"Why Systolic Architectures?"*. In IEEE Computer, pp.37-46, 1982, Volume 15.

Montague, R. *"Formal Philosophy: Selected Papers of Richard Montague"*. Edited by Thomason, Yale University Press, New Haven 1974.

Turner, D.A. *"Miranda: A non-strict, functional language with polymorphic types"*. Proceedings of the IFIP "International Conference on Functional Languages and Computer Architecture", Nancy (France), September 1985.

Chapter 3
GARBAGE COLLECTION

Programming languages for artificial intelligence applications, such as Lisp and Prolog, are required to handle large dynamic data or knowledge structures. Such structures require extensive support of algorithms and hardware to overcome memory limitations. Research in the area of garbage collection (GC) has been active since the late fifties when the first list–processing languages were implemented. A survey of GC algorithms is provided by Cohen (1981). The objective of GC is to reclaim memory space that is no longer used by the program; this activity is executed at run time and is usually transparent to the users.

GC algorithms can be implemented in software, hardware or a combination of both. However in order to obtain higher performance, speed and efficiency, dedicated hardware is usually provided.

GARBAGE COLLECTION FOR LISP SYSTEMS

Lisp performance is determined by the list processing speed and the efficiency of garbage collection (Hayashi, et al 1983).

Krueger (§3.1) proposes hardware support for *compacting, incremental* and *generational* garbage collection for a Lisp system. A modified RISC architecture (MIPS R2000) is chosen as the basis for implementing the GC support. Such an architecture implements read and write barriers with only a minor impact on conventional program execution.

GARBAGE COLLECTION FOR PROLOG SYSTEMS

Garbage collection is an important aspect in the implementation of any Prolog system (Appleby, et al 1988). The storage requirement grows during forward execution and is unwound during backtracking.

Bakkers, et al (§3.2) present an abstract machine, MALI, which owns and controls the dynamic memory space of a Prolog system. MALI has complete information about data structures and can, dynamically, detect any useless data and reclaim that memory space.

Other garbage collection strategies have been proposed, the reader is referred to Appleby, et al (1988) who describe a GC for the Warren's Prolog machine and Ridoux

(1987) who discusses a parallel approach.

References

Appleby, K., Carlsson, M., Haridi, S. and Sahlin, D., "Garbage Collection for Prolog Based on WAM," *Communications of the ACM,* vol. 31, no. 6, pp. 719–741, June 1988.

Cohen, J., "Garbage Collection of Linked Data Structures," *Computing Surveys,* vol. 13, no. 3, pp. 341–367, September 1981.

Hayashi, H., Hattori, A. and Akimoto, H., "ALPHA: A High–performance Lisp Machine Equipped with a New Stack Structure and Garbage Collection System," in *10th Int. Symp. on Computer Architecture,* pp. 342–348, 1983.

Ridoux, O., "Deterministic and Stochastic Modeling of Parallel Garbage Collection –Towards Real–Time Criteria," in *14th Int. Symp. on Computer Architecture,* pp. 128–136, 1987.

3.1 VLSI–APPROPRIATE GARBAGE COLLECTION SUPPORT

Steven Krueger

GARBAGE COLLECTION

Garbage collection is an important feature of the runtime support for some computer languages, especially for Lisp. In various applications a garbage collector may need to be compacting, incremental, generational, real-time or concurrent. For additional background on garbage collection, the reader is referred to Moon (1984).

A compacting collector coalesces holes in storage left by reclaimed data. Compacting requires copying the data that remains live after a collection and relocating references to it. Compacting storage simplifies the allocation of new storage and can improve the utilization and performance of backing (paging) store.

An incremental collector scans and copies a few words of data at a time, giving the appearance of never stopping the application. Baker (1978) developed the incremental, compacting collector. This type of collector was used in the MIT Lisp machine (Bawden *et al* 1979). Baker's collector uses a *read barrier* that traps attempts to read a data word which is a reference to storage existing from before this collection cycle. Preexisting storage is referred to as *oldspace*.

A generational collector uses the observation that much data becomes garbage shortly after its creation, or conversely, that data that has remained live after several collections is not very likely to become garbage on the next collection. Generational collectors group storage by age, collecting the newest objects frequently and collecting older generations at successively slower rates. Generational collection was first suggested by Lieberman and Hewitt (1980, 1983) and first used in the Berkeley Smalltalk system (Ungar 1984).

The key to generational collection is management of the references from older to younger generations. Each such reference is a root pointer, a reference to non-garbage in the younger generation. Data structures are needed that allow all references to a younger generation to be found without scanning all of the contents of the older generations. This bookkeeping requires a *write barrier* which traps stores of a reference to a younger generation into an older one.

A real-time collector has a definite bound on the time that garbage collection may add to the runtime of the application for any part of the application. In addition, the application with garbage collection must meet critical time requirements. Real-time garbage collection is not discussed further here.

A concurrent garbage collector performs all or part of the work of collection in parallel with the application. Concurrent collection will not be discussed further here.

The garbage collector used in the Compact Lisp Machine and in the Explorer computers is a compacting, incremental, generational collector which we call Temporal Garbage Collection (TGC). It has the further benefit of achieving high degree of memory locality and is described in Courts (1988).

GARBAGE COLLECTION SUPPORT

The principal requirement for supporting a compacting, incremental and generational garbage collection is the efficient implementation of the read and write barriers since they are exercised on every memory reference. The operation of these barriers is similar. Each is a complex function of the address and data of a memory transaction. Each always passes any data that is not a reference (eg. an integer).

On a read, the data is examined to determine if it is a reference into oldspace. The determination is made using a table which associates adddresses with the oldspace property. If the data refers to an address which has the oldspace property, the read is trapped so that the object may be copied.

On a write, if the data being written is a reference its generation is compared to the volatility of where it is being written. If the generation is lower than the volatility, the write is trapped so that table entries can be made in support of generational collection. Volatility indicates the youngest generation referenced from a page.

The logical functions of the read and write barriers are shown in Table 1.

Read Barrier		
	non-reference	reference
not oldspace	OK	OK
oldspace	OK	trap

Write Barrier		
	non-reference	reference
gen $\geq$ vol	OK	OK
gen $<$ vol	OK	trap

Table 1 Barrier Trap Logic

These barriers would be quite expensive without hardware support, requiring a read from a memory table and then several logical operations on each reference to garbage collected memory. Instead hardware barriers can work in parallel with other computation and interfere only when a trap is needed.

The architecture to implement the barriers is shown in Figure 1. The oldspace and generation maps are small RAMs. Page address translation may take several forms but is often an associative memory accessed by virtual page number. Page address translation also supplies the volatility which can be represented in 2 or 3 bits.

Other Uses of the Read Barrier

In the TI Explorer Lisp system, the read barrier has two other important uses. It is used to detect use of the unbound value and invisible data indirection.

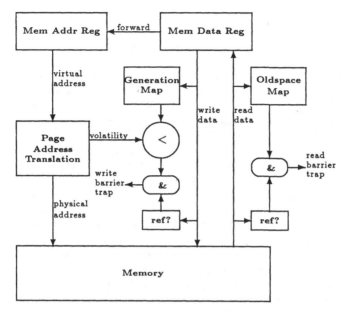

Figure 1 GC Barrier Architecture

Any use of the unbound value is caught by the read barrier and causes an error. The unbound value is given a special data type which always traps at the read barrier. Other illegal types are trapped at the read barrier too.

Invisible data indirection is used to implement a number of features of the Lisp system where it is usually termed *forwarding*. Array resizing, dynamic linking and dynamic closures are especially important features supported by data indirection. When an array grows it is not usually possible to allocate more space adjacent to its current location so a new, larger array is allocated and all references to the location of the old array are indirected to the new array.

When a Lisp function references another function, that reference is indirected through the location that contains the current definition of that function. Thus any use finds the current definition. This is termed dynamic linking.

A dynamic closure is a function and bindings for some dynamic variables. The Lisp system implements this by forwarding the current binding of the dynamic variables to the location of the corresponding value in the closure. This allows concurrent or nested invocations of the closure to share the same bindings with all having visibility to changes in the bound values.

Auto-Transporter Hardware

A circuit external to the Compact Lisp Machine processor chip (Bosshart *et al*, 1986) implements the barriers, forwarding and illegal type detection. When a memory read is started by the microprogram with auto-transport enabled, the oldspace map is consulted to obtain the oldspace property of the read data and its data type is used to index the transport decode table. This table contains two entries for each data type code, one is the transport action if the address portion of this word is a reference to oldspace; the other is the action if it is not. For non-reference types the two entries are the same.

Transport action may be TRAP, OK, or FOLLOW-FORWARD. The processor chip has a pin to accept the transport trap signal just as it has a pin to accept the address translation trap signal. A transport trap causes the microprogram to enter a trap handler to decode and act on the signal. All complex transporter actions begin by the auto-transporter signaling TRAP.

A transport action of OK signals the processor that the memory read operation has completed without trapping and the data is allowed to pass through the read barrier. This is the normal completion.

The transport action is FOLLOW-FORWARD when data indicating indirection has been read. The auto-transporter signals the processor which copies the data on the memory data pins to the memory address register and restarts the memory read with auto-transport. Because there is the possibility of very long or circular (hence infinite) chains of data indirection, the auto-transporter has a counter and signals TRAP instead of FOLLOW-FORWARDING when too many successive indirections have been encountered. The microprogram trap handler continues following indirections in an interruptible manner. We use a maximum of 7 indirections which is seldom encountered.

Similarly on write, the generation map and page address translator are consulted for the generation of the data and the volatility of the page where it is being written. The data type code is used as an index into a table with two actions per data type code, selected by whether the volatility is greater or equal to the generation. The action may be either OK or TRAP.

The auto-transporter interfaces with the processor chip as shown in Figure 2.

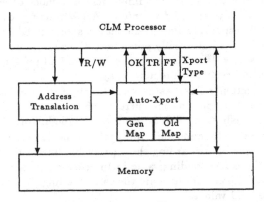

Figure 2 CLM Auto-Transporter

The auto-transporter is constructed with only a memory and some registers. On a read the output is determined by the data type of the memory data and whether the address part of the data word is a reference to oldspace. On a write the output depends on the data type of the data being written and whether volatility is larger than the generation. Our implementations have used RAM for flexibility but ROM could be used as well. The counter that detects when seven uninterrupted forwards have occurred is also implemented as a state machine in the RAM. The oldspace and generation maps are also RAM.

EXPERIENCE WITH GC SUPPORT HARDWARE

At Texas Instruments, we have been building GC support hardware in Explorer, Compact Lisp Machine and Explorer II since 1983. The complex, high performance garbage collection we have implemented is only possible with hardware support for the read and write barriers. In addition, the auto-transporter has increased system performance because it increased the performance of data indirection. In retrospect, it seems that the high relative performance of the auto-transporter on data indirection is mostly due to the low performance of the previous microcoded implementation rather than of any strong advantage in the architecture of the auto-transporter.

However, it is possible to implement some of the features requireing data indirection in other ways and to generally simplify the implementation. This leaner system with barrier detection support for garbage collection should give very good performance. Many changes to the Explorer software would be needed for such a system.

One other area where the auto-transporter could have been of greater utility is in support of TGC. TGC and the auto-transporter were designed concurrently and do not work well together. The particular mode of forwarding required by TGC is not supported by the auto-transporter. A considerable effort has been required to implement TGC with high performance in this environment.

APPLICATION TO A RISC MICROPROCESSOR

Steenkiste and Hennessy (1987) point out that a RISC microprocessor can perform well at symbolic computing and Lisp. This was the premise of the SOAR project at the University of California, Berkely (Ungar *et al*, 1984). Still, Lisp and other symbolic language implementations have not supported compacting, incremental, generational garbage collection with high performance, except on the high-level language architectures of the Lisp Machines. Therefore, a design sketch will be given of on-chip garbage collection barrier support for a conventional RISC microprocessor.

The MIPS R2000 is chosen for this modification because there is sufficient detail in the published literature to undertake this type of speculative design. The R2000 is well suited for this adaptation because it has on-chip address translation (Moussouris *et al* 1986) in the system coprocessor which can be easily extended to support volatility for the write barrier. The generation and oldspace maps and support logic will be on-chip as well.

The memory required for the generation and oldspace maps can be sized for the available silicon. If there are 4 generations, the two maps can be combined into a 3-bit wide memory which will be called the address space map (ASM). The granularity of these properties can be set so that the maps use the amount of RAM that can be constructed on-chip. For this design sketch, 96K bits of on-chip address space map will be used. This is large but practical. This size allows for 32K regions of memory. Since there are 512K pages in the user address space in the MIPS mapping scheme (DeMoney *et al* 1986), each region will have some integral number of blocks of 16 pages. For more frugal implementations the ASM can be sized to 48K bits or 24K bits with little harm.

Figure 3 shows various bit fields in the 32-bit process address. Bit 31 selects user or system space. Bits 1–0 select the byte within a word. Bits 30–16 are the region number used to index the address space map to produce oldspace and generation.

31	30 16	15 12	11 2	1 0
sys usr				byte
	region number			
	page number		page offset	

Figure 3 Address Usage

The address space map is added to the R2000 by routing virtual address bits 30–16 to a 32Kx3 bit RAM. We do not need garbage collected memory in system space. To detect errors, any reference using the barriers will trap if the data is a reference to system space.

It is important that the barrier support hardware introduce no delay when it is not needed, such as in a program that isn't using garbage collection, when getting a character from a string or when fetching an instruction. Since the read barrier involves processing after the data has returned from the cache, the barrier may delay the time when data is first usable.

A study of the barriers in relationship to the other pipeline activities in the R2000 reveals what is possible. A pipeline timing diagram based on Figure 3 of Moussouris *et al* (1986) is shown in Figure 4. The read and write barriers are assigned a pipeline stage at the earliest time that all inputs are available. Each is assigned the same amount of time as an address generation/translation stage since the computation of a barrier is similar, involving a read of an internal RAM and a few gate delays of logic.

The dotted line in the figure shows when the read barrier of instruction 0 completes and cuts through the other instructions in their various stages of progress. As none of the following instructions have written results to registers (the write back stage), all are abortable so that the read barrier trap happens cleanly with no instructions executed beyond instruction 0. Only a store to memory in instruction 1 would have proceeded too far to abort since the memory is written in the data cache stage. But a store is idempotent as long as it is to a different location than the load at instruction 0. So to remain restartable, a load using the barrier must not be followed immediately by a write to the same location.

The write barrier timing is more favorable as all of the data for the operation of the write barrier is available at the end of the data address stage. The write barrier stage is shown in Figure 4 on instruction 1 only. The write barrier completes after the memory is written in the data cache stage. This is not a problem because none of the write barrier trap handlers require the previous contents of the location.

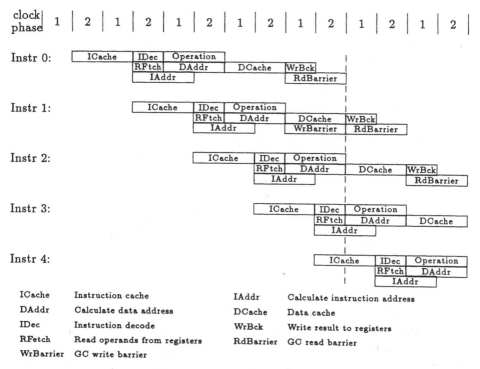

ICache — Instruction cache   IAddr — Calculate instruction address
DAddr — Calculate data address   DCache — Data cache
IDec — Instruction decode   WrBck — Write result to registers
RFetch — Read operands from registers   RdBarrier — GC read barrier
WrBarrier — GC write barrier

Figure 4 Barrier Pipeline Timing

Typing

As has become common in Lisp systems for conventional hardware (Steenkiste 1987), data typing is encoded in the low order two bits of a data word. This is the part of an address that specifies the byte within a word. All typed accesses are word aligned so that there is no need to specify a byte within the word. The R2000 will trap to indicate an attempt to use an address that is not properly aligned for the data size of memory operation. All barrier reads and writes are full word operations, so addresses without 0 in the low order two bits cause the unaligned address trap. The unaligned address trap can be used for type checking at low cost.

The data type is encoded so that a 1 in bit 1 indicates a reference and a 0 indicates a data value that is not a reference. This encoding allows for very simple detection of references in the barrier logic. The type bits are 3 for a list. Types 0 and 1 are for data values contained completely in the rest of the data word. Integers are type 0. This choice allows two integers to be added together without the need to mask or restore the type codes. Type 1 is for character. All other kinds of data are represented as type 2, structure. Structure is the other reference type. The exact type of the structure referenced is indicated in the memory it references.

Memory address calculations generated by the compiler for access to list and structure data will need to perform alignment. Address calculations for list data should be compensated by subtracting 3. Address calculations for structure data should be compensated by subtracting 2. To access the third word of a structure the offset would be $3 \times 4 - 2 = 10$, the third word times 4 bytes per word minus 2 to align a structure reference. If the value used as the base for this reference is not a structure reference, an alignment trap will be raised.

Multiple Address Spaces

Notice that there are no provisions for sharing the address space map among several process address spaces. This would use additional chip area to store process ID and also degrade performance severely. The scheme proposed requires that every virtual address that is referenced have valid generation and oldspace map entries. The scheme can be changed to fault on the use of invalid entries, but in our previous implementation on Explorer this decreased performance by several percent in a single address space. With multiple address spaces contending for map entries and the wide address distribution of objects referenced in Lisp, many reloads may be required.

New Instructions

The existing R2000 instructions do not use the barriers and are not changed. Instead new load and store instructions have been added to the R2000's instruction set. This allows existing compilers to be used for conventional languages and prevents trapping in programs without need for garbage collection and on references to untyped data in Lisp programs. The new instructions are LWGC (Load Word with Garbage Collection) and SWGC (Store Word with Garbage Collection). These are the same as LW (Load Word) and SW (Store Word) respectively, but trap on barrier violations. The new instructions are full word operations and trap on alignment violations.

System Coprocessor Support

In order to support the processing of barrier traps, two new control registers need to be added to the system coprocessor. In addition, the existing control registers (see DeMoney *et al* 1986) are utilized on a barrier fault. The trap sets the Cause register to indicate a read or write barrier fault in the *ExCode* field, which is widened to 6 bits to accommodate the new exceptions. The Exception PC register is set to the instruction following the load or store on a barrier fault. The Bad Virtual Address register is set the memory address that was referenced to get the barrier trap. The data can be reread from that location on a read barrier trap and also on a write barrier trap since the write completes before trapping.

The system coprocessor also needs instructions and registers to access the address space map. The access registers are shown in Figure 5. The AS Map register is new and gives access to the generation and oldspace bits in the address space map. The Entry Lo register existed in DeMoney *et al* (1986) to access the TLB entries. It has been augmented with the 2-bit *Vol* (volatility) entry. The AS Map Index register contains the portion of an address to access the address space map.

Two new system coprocessor instructions give access to the AS Map register. They

AS Map Index:

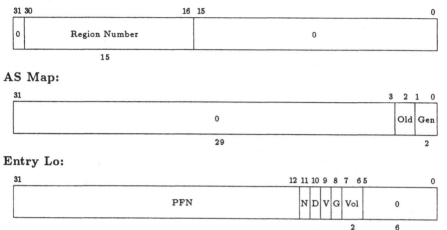

Figure 5 Address Space Map Access Register

are Read Indexed AS Map and Write Indexed AS Map. Read Indexed AS Map reads the address space map entry addressed by the AS Map Index register into the AS Map register. Similarly, the Write Indexed AS Map instruction writes the address space map entry addressed by the AS Map Index register with the contents of the AS Map register.

In servicing a barrier trap, the AS Map Index register is loaded by reading from the memory location addressed by the Bad Virtual Address register. Then a Read Indexed AS Map instruction will bring the relevant information from the address space map into the AS Map register where it can be used in processing the fault.

CONCLUSION

The experience gained from the garbage collection hardware for the Compact Lisp Machine can be applied to more conventional processor architecture. Even a RISC microprocessor can be augmented for sophisticated garbage collection and the additional hardware can be added in a compatible way with little impact on the overall architecture. The resulting integrated circuit is buildable in today's technology.

ACKNOWLEDGMENTS

This paper would not have been possible without the considerable work in garbage collection and symbolic computer architecture at Texas Instruments over a number of years. Bob Courts and John Osman created TGC and conducted many measurements on Explorer systems. Pat Bosshart conceived of the auto-transporter. Arthur Altman, David Bartley and Don Oxley showed me the promise of the RISC methodology for symbolic processing.

Explorer and Explorer II are trademarks of Texas Instruments Incorporated. The Compact Lisp Machine was supported by US Navy contract N00039-84-C-0537 and Texas Instruments.

References

Baker H, "List Processing in Real Time on a Serial Computer," *Communications of the ACM*, vol. 21, pp. 280–294, 1978.

Bawden A, Greenblatt R, Holloway J, Knight T, Moon D, and Weinreb D, "The LISP Machine," in *Artificial Intelligence: An MIT Perspective*, P Winston and R Brown (ed), Cambridge, Mass: MIT Press, pp. 343–373, 1979.

Bosshart P, Hewes C, Chang M, Chau K, Hoac C, Houston T, Kalyan V, Lusky S, Mahant-Shetti S, Matzke D, Ruparel K, Shaw C, Sridhar T and Stark D, "A 553K-Transistor LISP Processor Chip," in *Int. Solid-State Circuits Conf. Digest of Technical Papers*, pp. 202–203, 402, 1987.

Courts R, "Improving Locality of Reference in a Garbage-Collecting Memory Management System," *Communications of the ACM*, vol. 31:9, pp. 1128–1138, 1988.

DeMoney M, Moore J and Masey J, "Operating System Support on a RISC," in *Digest of Papers Spring COMPCON 86*, San Francisco, Calif, pp. 138–143, 1986.

Lieberman H and Hewitt C, "A Real Time Garbage Collector that can Recover Temporary Storage Quickly," Technical Report 569, Artificial Intelligence Laboratory MIT, 1980.

Lieberman H and Hewitt C, "A Real Time Garbage Collector Based on the Lifetimes of Objects," *Communications of the ACM*, vol. 26, pp. 419–429, 1983.

Moon D, "Garbage Collection in a Large Lisp System," in *Conf. Record of the 1984 ACM Symp. on LISP and Functional Prog.*, Austin, Texas, pp. 235–246, 1984.

Moussouris J, Crudele L, Freitas D, Hansen C, Hudson E, March R, Przybylski S, Riordan T, Rowen C and Van't Hof D, "A CMOS RISC Processor with Integrated System Functions," in *Digest of Papers Spring COMPCON 86*, San Francisco, Calif, pp. 126–131, 1986.

Steenkiste P and Hennessy J, "Tags and Type Checking in LISP: Hardware and Software Approaches," in *Proc. of Second International Conf. on Architectural Support for Prog. Languages and Operating Systems*, Palo Alto, Calif, pp. 50–59, 1987.

Ungar D, "Generation Scavenging: A Non-disruptive High Performance Storage Reclamation Algorithm," in *Proc. of the ACM SIGSOFT/SIGPLAN Software Engineering Symp. on Practical Software Development Environments*, Pittsburgh, Penn, pp. 157–167, 1984.

Ungar D, Blau R, Foley P, Samples D and Patterson D, "Architecture of SOAR: Smalltalk on a RISC," in *Proc. 11th Annual Int. Symp. on Computer Architecture*, Ann Arbor, Mich, pp. 188–197, 1984.

3.2 A SELF–TIMED CIRCUIT FOR A PROLOG MACHINE

Yves Bekkers, Louis Chevallier, Serge Le Huitouze, Olivier Ridoux

INTRODUCTION

An abstract memory, called MALI: "Mémoire Adaptée aux Langages Indéterministes", has been designed at IRISA. The storage function of this memory is well suited for implementing relational non-deterministic languages such as Prolog (Bekkers *et al* 1984, 1986). A bi-processor Prolog-machine has been designed on the basis of this abstract memory: the main processor carries out system calls and performs the Prolog interpretation while a dedicated processor is intended for memory management. We describe the circuit which is the heart of the memory processor. The global architecture of the machine and the architecture of this circuit are presented.

PROLOG and the MALI PROJECT

Prolog implementations suffer from two problems of high level languages: execution speed and memory consumption.

Compilation generally achieves fairly good speed performance (comparable with the efficiency of other high level languages). Dedicated hardware is another means to increase significantly the number of Logical inferences per Seconf (LIPS).

On the other hand, the memory consumption problem is only partially solved. In order to save memory, regular implementations (e.g. Warren's machine) resort to both static and dynamic solutions: static partition of variables into local and global classes, tail recursion optimization, recovering of stacks space upon backtracking and garbage collecting of the global stack. In spite of these efforts, none of the currently available Prolog interpreters implements a complete memory recovering. Hence, for any Prolog systems, there always exists a program whose execution leads to memory exhaustion even when this program actually needs a finite or fair amount of space.

With MALI, dynamic objects, including the data structures needed for *Or*-control, are gathered in a specific place. The management of this pool is up to MALI.

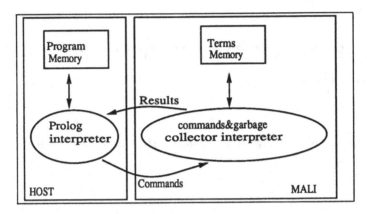

Figure 1 Architecture of a Prolog/MALI system

Altering and consulting must be done through MALI. Inside its memory, information, which is a set of logical *terms*, is represented with an homogeneous set of linked cells.

Our algorithm achieves complete recovery of unused information representation inside MALI's memory. This result is obtained because MALI is in a position to take into account the *non-determinism* and hence to deduce the precise state of the resolution.

A suitable collection algorithm (Baker 1978) allows parallelism between the process of interpreting Prolog and the collecting process.

MALI — THE ARCHITECTURE

The global architecture is composed (Figure 1) of a host which carries out interpretation of the Prolog programs and supports the operating system and of MALI which provides services requested by the host. MALI takes charge of dynamic objects needed during the interpretation. The host obtains from or delivers information to MALI via a (small) shared memory. Each transaction goes through with an asynchronous protocol and carries a service identifier. The services that MALI provides relieve the host from any memory management burden and, especially, they recognize, among other commands, "save" and "retrieve", the instructions which are the basis of the simulated non-determinism. To perform a unification the service "substitute" is available; it requires the name of a *variable* and the term it has to be bound to.

With MALI, unlike most others abstract machines, the object *variable* is a "true" variable that has to be free in order to be visible by the host. Once a variable has been bound, it simply disappears; it will possibly come back upon backtracking.

The MALI processor supports two tasks:

- Servicing the host requests which include build terms and retrieve them, and save or cancel a *state* or backtrack to the previous.

- Collecting the useful cells and getting rid of the others.

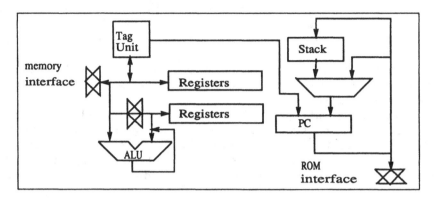

Figure 2 Internal architecture

There are three distinct memories in the machine:

- The host memory, where the Prolog program resides.

- The small common memory between the host processor and MALI.

- A large memory, owned by MALI where the terms are stored.

At the present time, three implementations of MALI exist: a software version and two hardware boards: one with a Multibus interface and the other for IBM PC compatible bus. These boards has been built around the AMD29116 and the AMD2910.

Hardware boards allow real parallelism to be obtained, a VLSI circuit is under design. Actually, we need more reliable, less cumbersome and more efficient machine for installing it into personal computers, since the target is to build a cheap Prolog machine from any commercial personal computer.

THE CIRCUIT

Internal architecture

As in many symbolic machines, MALI uses tagged words (cells). In order to retain uniform cells size, a tag does not refer to the cell it is in, but to the cell it points to. The cells are 32 bits long, the tag comprises up to 8 bits, the rest is for information (atoms or pointers).

The circuit is a microprogrammable processor. The microprogram that resides in an external ROM, contains the code for both services and collecting algorithm.

Servicing a user request consists of the execution of some instructions. Upon a request from the host, the background task (the garbage collection) is generally suspended immediately and resumed later. However the collecting algorithm has some non-interruptible sections which may delay the acknowledgment.

Inside the circuit (Figure 2) are two 16-register files; these registers are 32-bits wide. They serve as working registers for the services and collection algorithm. A 32-bit adder performs address calculation. A tag-unit is able to quickly compare, synthesize and test tags.

A sequencer is made up of an 8-words deep and 16-bits wide hardware stack, a 16-bit program counter and a multiplexer which enables multi-way branches according to bit fields taken from the tag.

For the sake of performance, MALI must be able to respond very quickly to any service request. To achieve this, we want non-interruptible subsections to be as short as possible, and unique resources, such as accumulators and various temporary registers, are therefore undesirable. The instruction-set provides register-to-register and register-to-memory type instructions. Only two addressing modes are available: absolute and immediate. It is possible to affect the interruptability on each branch-instruction.

The set of registers is split in two files. One of them is devoted to address handling. This one is two-ported (two registers may be read at a time). The second file has a more general purpose; it has one port and its size can be extended according to the place available on the die. One can access directly the 16 first registers while indirect access is needed to use any registers. Furthermore, these registers provide an internal memory accessible by the host through the host interface.

The terms memory interface and the host interface share the same 32-bits wide datapath through which commands identifiers, parameters and results are passed. Since many services turn out to be simple memory accesses the host interface includes signals which make possible to perform this memory access while the address is tested.

Clocking

The clocking of this circuit does not follow the classical synchronous methodology. The circuit spends much of its time accessing shared memory. Parameters and results of commands go through a memory which is shared between MALI and the host. Therefore, the interface with these memories have to be asynchronous.

We have decided to employ a *self-timed* strategy (Seitz 1980). This discipline inherently fits well with asynchronous interface problems. Moreover, one can see that our circuit has two roles: servicing host requests and performing memory collection. Hence services have to be implemented as interrupts, but, in this case, they have to be accepted very quickly. Performances are globally improved with self-timed clocking because it yields a better synchronization between the host and MALI. Besides, the moderate size of our circuit makes it possible to bear the complexity overhead introduced by self-timed policy.

The self-timed policy consists of interconnecting modules, each module being bound to follow a four-phase-protocol shown in Figure 3. The interconnection is committed not to violate this protocol for each modules. No clock is needed: the modules must signal the completion of their own phases, this signal is fed back in the network and serves as request or acknowledge for other modules. In the Figure 4 the two modules infinitely run through their two phases. A special element (the C-element) forces them to synchronize twice a cycle.

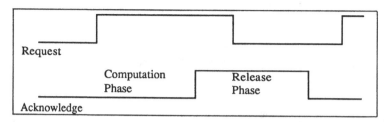

Figure 3 The four phases cycle

We represent the simplest modules with a box. It has one input and one output, which carry, respectively, the request and the acknowledgement. A module is free to delay its responses for an arbitrary amount of time but it must eventually do it.

The level a device is self-timed is defined by the size of the simplest modules (Clarke 1973), i.e. modules which cannot been decomposed further into smaller modules.

We set the modules to be as complex as the file registers, the address generator, the hardware stack, the instruction decoder (a PLA) or the program counter. There are a dozen main modules inside our circuit. The memories interfaces are themselves self-timed modules while they are outside the circuit.

Other self-timed designs have chosen bit-level modules. While this yields a greater potential parallelism between the modules activity thanks to a very fine grain synchronization, it also incures a very high complexity overhead.

We have chosen a three-level pipeline implementation whose stages are:

- fetching micro-instructions;

- decoding, fetching operands (from registers) and storing results of the previous instructions and

- executing them.

Of course, the pipeline synchronization is performed in a self-timed manner. The registers between the stages form new self-timed modules (Hayes 83). The potential parallelism between the three previously tasks can be naturally expressed and wholy obtained by using self-timing.

Circuit Considerations

In order to implement the circuit some special devices are needed however; the C-element or Muller-element which computes the rendezvous between two signals is of central importance. For example, at the top-level of the architecture of the circuit a rendezvous must be accomplished between the three stages at each cycle. This element is a sequential gate which toggles when the value of all of its inputs become the same.

In order to improve the output of the machine it desirable to provide an instructions buffer so that the next instruction can be fetched in advance so that whenever

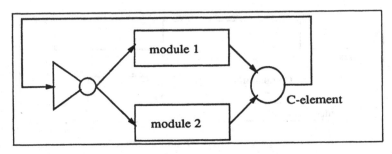

Figure 4

the execution stage is late (perhaps waiting for a memory access) no delay is lost fetching instruction memory. This buffer is only one instruction long.

The circuit should exhibit the two properties of liveness and safety. The first one means the circuit is bound to never get stucked in some state, the second one states that, given a initial state, no erroneous configurations can happen. Such configurations would be, for example, a latch with its two gates opened, so that data can flow through it without any memorization.

To obtain the oscillation we need some non-linear devices inside the loop, classical inverters do have this behavior but we employ Schmitt-triggers instead. They are designed to convert any input level to a "logic" one, and we can make use of this feature to detect the completion of modules.

The unavoidable four-phase protocol would be less cumbersome if we could put some useful work into each of its phases. In many cases, while the actual computation takes place within the compute-phase, we can take advantage of the release-phase to precharge some electric nodes (busses) or to evaluate another part of the module. These tricks depends on how the logic function of the module is implemented.

In Figure 4 the modules have to make a rendezvous twice a cycle. Since this method may waste time one can lessen the coupling between the modules with the circuit of Figure 5. This circuit still exhibits liveness, and still enforces the four-phase-cycle for both modules.

MALI is both a co-processor with regard to the host and a autonomous processor when achieving collecting, so it has to service requests as quickly as possible while performing its background task. We can take advantage of self-timing for the necessary synchronizations with the host: the well-known metastability problem which plagues synchronous designs is gracefully overcome by self-timed designs: we simply use a module which is able to realize when its output has settled in a logic level. We no longer need to waste a delay, on each synchronization event, to avoid the metastability risk.

To detect the completion of the computation of non-compound modules one can resort to combinatorial or electrical means. A last solution is to use a fixed delay which is longer than the known slowest response of the module.

The ALU offers an opportunity to combinatorially detect the end of its activity: at the center of it, is a carry chain (the way it is implemented is no relevant). Designers

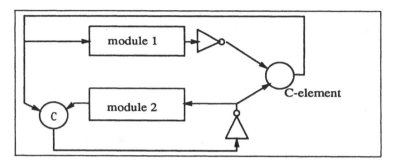

Figure 5 Modules interconnected with a less tight coupling

strive to shorten the duration of the carry propagation from the bottom to the top of the chain, because, in a synchronous design, the worse case has to be considered, even if it almost never occurs. We are using two carry chains, they concurrently evaluate the carries and their logical complement. Given the initial state of all ones on both chains, the work is done as soon as all carries are pairwise different. The average computation time is the average length of the longest uninterrupted propagation carry sub-chain, 5 by 32 bits words.

Completion detection of the evaluation of the PLA, read and write operations in registers are performed with level sensing.

Verification of the design

Temporal logic is a well established formalism useful to specify constraints on signals inside our self-timed modules networks (Dill 1985, Martin 1985, Malachi 1981). We use a temporal logic simulator to validate successive designs of the machine. Indeed, classical logic simulators do not provide a suitable model for time as they only carry out a deterministic analysis of possible executions. Now each module can take an arbitrary amount of time for completing their task, this delay is even less foreseeable when this module is a compound module.

Since any module can be described with a set of temporal formulae, the whole system could be modeled in the same way giving the formula f. The constraints are themselves other formulae (c), then the verification could merely be the proof of the validity of f $\Rightarrow$ c. Unfortunately, this approach is generally unworkable because of the non-polynomial complexity of the demonstration. The classical solution is to build a graph which represents the whole model of f, then constraints formulae are compared with the graph (Dill 1985) whole model of f, then constraints formulae are [DC85].

CONCLUSION

An abstract memory, MALI, has been presented. This memory is the basis of the bi-processor Prolog-machine we are building. We have then described a VLSI circuit intended for implementing MALI. The distinctive feature of this circuit is a self-timed clocking which we feel well suited for its double job: a co–processor for servicing of MALI commands and an autonomous processor for collecting memory.

References

Baker, H. G., "List Processing in Real Time on Serial Computers," *Comm. of the ACM*, vol. 21, no. 4,   April 1978.

Bekkers, Y., Canet, B., Ridoux, O. and Ungaro, L., "A Memory Management Machine for Prolog Interpreters," in *Proc. 2nd Int. Logic Programming Conference*, Uppsale, Sweden, 1984.

Bekkers, Y., Canet, B., Ridoux, O. and Ungaro, L., "Mali: A Memory with Real-time Garbage Collector for Implementing Logic Programming Languages," in *3rd Symposium on Logic Programming*, Salt Lake City, 1986.

Clarke, E., Emerson, E., and Sista, A., "Automatic Verification of Finite State Concurrent Systems using Temporal Logic Specifications: A Practical Approach," in *10th ACM Symp. on Principles of Programming Languages*, , 1983.

Dill. D. and Clarke, E., "Automatic Verification of Asynchronous Circuits using Temporal Logic," in *Chapel Hill Conference on VLSI*, Chapel Hill, North Carolina, 1985.

Hayes, E., "Self–timed IC Design with PPL's," in *3th Caltech Conference on VLSI*, Pasadena, Calif., March, 1983.

Maritn, A., "The Design of a Self–timed Circuit for Distributed Mutual Exclusion," in *Chapel Hill Conference on VLSI*, Chapel Hill, North Carolina, 1985.

Malachi, Y. and Owicki, S., "Temporal Specifications of Self–timed Systems," *VLSI Systems and Computations*, 1981.

Seitz, C., "System Timing," in *Introduction to VLSI Systems*, Mead, C. and Conway, L., Addison–Wesley, 1980.

Chapter 4

CONTENT–ADDRESSABLE MEMORY

A content–addressable memory (CAM) allows parallel access of multiple memory words; CAM has an inherent parallelism that can have an impact on the performance of some architectures for artificial intelligence applications. Parallel search and parallel comparison are the major advantages of CAMs over random access memories. CAM memory cells have in addition to the memory circuitry a comparison circuit for pattern match. Although CAMs are not new (a survey is provided in Yau and Fung 1977), VLSI technology has made possible to built static and dynamic CAM at a relatively low cost.

CAMs have been used in SIMD architectures such as STARAN (Batcher 1977), OMEN (Thurber 1976), RELACS (Berra and Oliver 1979) and WASP (Lea 1986). However, little has been done on using these memories for architectures for production systems and logic programming.

CAM FOR AI SYSTEMS

Content–addressable memories can provide fast parallel search and matching, which are computationally intensive, for production systems and logic programming. In this chapter three approaches are presented.

Kogge *et al* §4.1 present a hardware support for logic rule–based programming systems. A 64×32–bit CAM chip is briefly described. Basic algorithms for handling AI structures like lists, stack and list comparison are described. Performance on Prolog execution and production system computation can be improved by CAM system; a performance analysis is shown in the paper. Prolog support is provided in three areas: *clause filtering* to determine which clauses in the program are useful for the current goal; *unification* to support retrieving and storing results of the binding operations; and *stack management* to support fast backtrack and dereferencing operations. Rule–based systems that have been compiled into Rete Nets or equivalent networks are good candidates for CAM–based systems.

Another system based on CAM is described by Ng *et al* §4.2 who focus their investigation around the unification function. The authors study the unification requirements and data organization in order to propose a new processor architecture. The VLSI implementation of the CAM is based on the Wade/Sodini cell and it is proposed to have 100×40–bit words.

Robinson §4.3 describes an array system that is based on an associative memory integrated circuit; the chip has a capacity of 1152×20–bit words. Parallelism is exploited on-chip and at the system level; this yields a high system performance. Pattern matching is the system's main function: data stored in the pattern address-able memory (PAM) is matched with the input query data. The PAM described here can handle arbitrary length strings of symbols. Functions to handle I/O and perform modifications on selected stored data are built–in.

References

Batche·, K. E., "The Multi–dimensional Access Memory in STARAN," *IEEE Trans. on Computers,* vol. C–29, no. 9, pp. 836-840, Sept., 1977.

Berra, P. B. and Oliver, E., "The Role of Associative Array Processors in Database Machine Architecture," *IEEE Computer,* vol. 12 no. 3, pp. 53–61, March, 1979.

Lea, R. M., "WASP: A WSI Associative String Processor for Structured Data Pro-cessing," in *Wafer Scale Integration,* C. R. Jesshope and W. R. Moore (Eds), Bristol: Adam Hilger, pp. 140–147, 1986.

Thurber, K. J., *Large Scale Computer Architecture—Parallel and Associative Proces-sors.* N.J.: Hayden Book Co., 1976.

Yau, S. S. and Fung, H. S., "Associative Processor Architecture—A Survey," *ACM Computing Surveys,* vol. 9, no. 1, pp. 3–28, March, 1977.

4.1 VLSI AND RULE–BASED SYSTEMS

Peter Kogge, John Oldfield, Mark Brule and Charles Stormon

INTRODUCTION

Many AI applications today employ some form of *if–then* rule–based programming. In languages stressing *production rules*, such as OPS, data is pattern matched against the *if* part until all literals are satisfied; the *then* part then indicates how the database should be changed. In more *deductive* languages like Prolog, the match is between a goal literal whose truth is unknown, and the *then* part of a rule. A successful match causes a series of goals from the *if* part to be solved in a similar manner.

This paper addresses an accelerator for such languages: a VLSI coprocessor designed around a *Content Addressable Memory* (CAM). Usefulness of such a device will be demonstrated in three areas of rule–based language processing: determining which rules may be applicable, pattern matching, and overall sequencing. Data from the well–known "Monkey and Bananas" benchmark will be used to demonstrate effectiveness.

AN AI–ORIENTED CONTENT–ADDRESSABLE MEMORY

Content–addressable or Associative Memories are not new, but have become more useful as MOS memory technology has improved. Both static and dynamic basic CAM cell circuits exist. For example, by adding a simple comparison circuit, the classic 6–transistor static RAM becomes a 9–transistor static CAM cell. Other extensions give the ability to store a third "don't care" state as well as "1" or "0" (called a *trit*).

When configured into chips, CAMs can be built with a conventional address decoder to select individual word rows, but for AI applications it is valuable to base row selection on the results of previous search operations. This can be obtained by adding a small data flow to each word (*row logic*). In effect, the CAM becomes more than a memory – it is a SIMD computer.

Over the last four years we have investigated at Syracuse University three generations of CAM chips optimized for AI applications (Oldfield *et al* 1987). Figure 1 represents the floorplan of the most recent chip. It uses a ten–transistor static CMOS cell, with two bit cells per trit used when needed. (We have developed a new cell for static trit storage which requires fifteen transistors that will be reported in the future). In MOSIS 3–micron CMOS design rules one of these cells occupies 60μ x 57μ, which compares well to the 5 transistor 52.5μ x 44μ

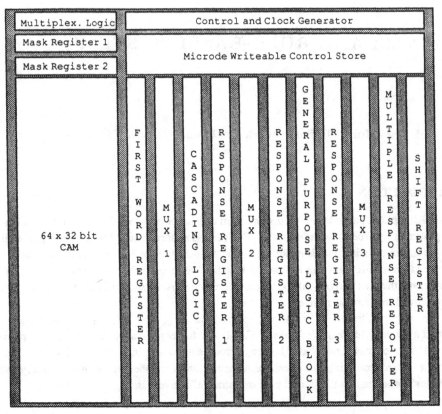

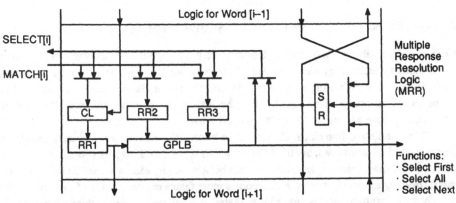

RR1, RR2, RR3, SR = 1 bit Latches
GPLB = General Purpose Logic Block = Any of 256 Logic Function of 3 Inputs
CL = Select Either MUX Output or RR1[i–1]

Figure 1. The Syracuse University CAM Chip

Sodini and Wade (1987) dynamic cell in the same technology. Typical search times are in the 50 ns range, and 64 rows of 32 cells each have been integrated on a single chip of overall dimensions 9200μ x 7900μ, including all of the row logic, sense amplifiers, and control circuits.

Search operations are carried out in parallel, using a match line passing through every cell of each word. All match lines are precharged high at the start of a search operation, and any cell whose contents do not match the bit presented on the corresponding column data lines will discharge the match line to ground. Search patterns may include "don't cares" as well as 1 or 0. This is accomplished by storing the appropriate "care"/"don't care" pattern in a mask register.

Cascading CAMs is important, both in terms of adding more words and in increasing word length. In our CAM the physical words are 32 bits in length, and the chip architecture permits cascading chips to increase the number of words. For increasing effective word length there is an additional CAM bit for each physical word (called the *First Word Register* or *FWR*), (Stormon *et al* 1988), as suggested by Lee and Paul (1963). When a data object is written to CAM the FWR is set to "1" for the first 32 bit physical word and "0" for all remaining words. This allows the logical word size to be any integral multiple of 32 bits up to 2K bits.

Row Logic

Much of the suitability for AI applications in this chip comes from the row logic attached to each word. This is a robust one bit data flow containing several one bit *Response Register (RR)* latches, some multiplexers, a *General Purpose Logic Block (GPLB)*, and interfacing to the row logic for neighboring cells. Coupling to the associated CAM word is through a *Match* signal that indicates the match result between that CAM word and the input, and a *Select* signal which selects the word during CAM updating. These signals pass through the FWR cell, permitting it to take part in the row logic's decision making process. The GPLB can perform any Boolean function on its three inputs, with its result usable throughout the row logic. A single set of control lines drives all row logics simultaneously.

The *Multiple Response Resolution Logic (MRR)* handles the multiple results that a single matching cycle may provide. Three modes of operation are provided, selected by means of control lines. The first is the standard multiple response resolution, *Select First*, in which only the topmost active row has its output selected. The second mode activates not only the topmost output but also all the outputs below it (*Select All Below*). Finally, the third mode, known as the *Select Next*, activates the output of the next active input which follows the topmost selection and resets the topmost. The *Shift Register (SR)* permits the outputs of the MRR to be shifted up and down the CAM array.

SOME IMPORTANT CAM ALGORITHMS

Multiword Entries

In implementing real AI programs, objects often require more than 32 bits. The FWR bit in our CAM can be used to signal the start of a multiword object. Assuming that there are many such multiword entries in a CAM, and we wish to see which one(s) match some other multi-word object (outside the CAM), the following algorithm would perform the compare in a time proportional to the number of 32 bit words in the object:

1. Compare the object's first word to all CAM words whose FWR = "1".
2. Shift the result of the compare down to the next word (via SR logic).
3. Compare the second word to all CAM words selected by this shifted value.
4. Shift the result of this compare down one more, and repeat as needed.

Stack Management

Assume we wish to keep a stack in CAM, with each stack frame an unpredictable length. Again assume that the stack is built from consecutive CAM words, bottom up, with the topmost word of each frame indicated by FWR="1".

Pushing a new word to the topmost stack frame involves finding that word (a single CAM operation with the MRR function set to "Select First"), and using the shift logic to select the next highest cell for the write. This same path can then be used to reset the prior topmost FWR, and set that for the new one.

Popping a frame involves searching the CAM for all words with FWR set, and then using the MRR to "Select Next". A "Select All" using this response identifies all words associated with the stack below the top one. A third operation then ANDs this indication into the FWR, clearing that for the old top.

There are times when each stack frame is a collection of multiword entries itself. In such cases one of the RR latches can hold the "top of stack" entry, leaving the FWR for its original first word indication.

List Comparison

Central to most AI processing is the need to perform sophisticated comparisons against complex objects such as lists or trees. In conventional computers such objects are linked lists, with virtually all operations requiring tedious pointer following from a root to a leaf.

With a CAM like ours, such objects can be represented in a compacted form where only the leaves are recorded (Sohi *et al* 1985). This compaction tags each leaf by a binary representation of its position in the list, as pictured in Figure 2 with a list of standard operations. More complex operations such as list equality are possible by comparing the leaves of one list (with tags included) one at a time against the CAM entries.

SUPPORT FOR PROLOG

Figure 3(a) diagrams a typical view of how a Prolog program operates. The key data structure is a stack of frames, one per inference, containing a list of bindings of values to variables made in that inference, a list of subgoals left to be proven, and an indication for that goal what clauses are left to be tried. Computation consists of looking for a possible clause to satisfy the

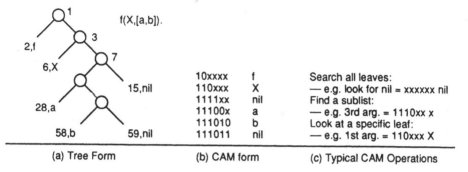

| (a) Tree Form | (b) CAM form | (c) Typical CAM Operations |

Figure 2. List Comparison with CAMs

top goal on the stack, attempting to unify it with the head of some rule, and then modifying the stack for the next cycle.

All of these steps can be improved by use of a CAM. We have implemented a Prolog interpreter for Figure 3(b) where the CAM supports the stack of binding entries and represents the head literals of the rules. Simulations of the CAM operations were done at the bit level on a Connection Machine (built by Thinking Machines, Inc.) at Syracuse University.

Filtering

The number of attempted unifications of goals against rule heads can be reduced by performing a preliminary *filtering* function. The approach used here is derived from the idea of superimposed code words (Colomb 1986), and relies on the CAM's ability to store and process "don't care" values, as well as handle objects of multiple and varying sizes.

Whenever a group of clauses with the same head predicate are encountered in a program, they are examined as a unit, and the arguments analyzed for which ones actually distinguish one rule from another. For example, in the following, only the first and third arguments provide distinctive information; the second argument is a variable in all clauses, and will thus match any goal term.

append(nil,L,L).
append([H|T],L,[H|Z]):–append(T,L,Z).

Once the block of rules has been so analyzed, a *code word* for each rule head is created. Each argument is encoded into a 15 bit field, using encoded forms of the value if it is a ground term (such as *nil*), or "don't care" when it is specified in terms of variables (as with L). These 15 bit fields are concatenated to form a single code word. If the code word size exceeds 96 bits, the remaining information is superimposed by logically OR–ing into the first 96 bits, starting at the beginning of the code word. The predicate's name and arity, as well as the database index, are stored in an additional 32 bits, creating a 128–bit entry in CAM for each clause in the program.

When a goal is presented for resolution, the arguments are *dereferenced* (variables replaced by their values) to the point where equivalent 15 bit code fields can be created. Once

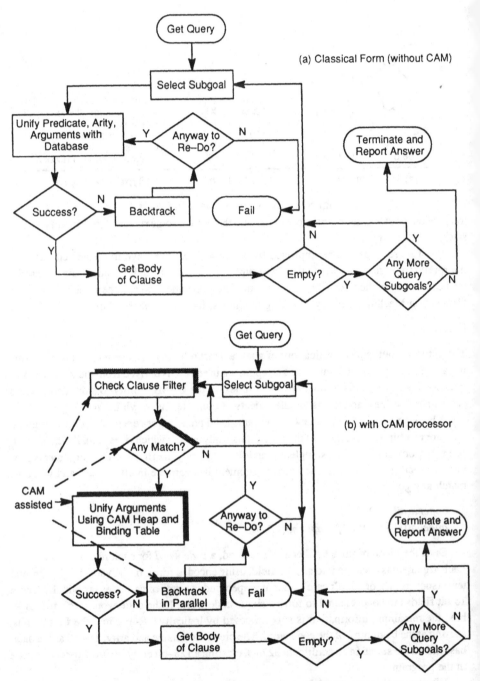

Figure 3. Use of CAM in Prolog Execution

again, free variables are encoded with "don't cares". These fields are used to create the 128–bit code word as for the rule heads, and the resulting code word used in a CAM match. Any match signals a rule that might work with the current goal. All other rules are known to be irrelevant, and need not be tried.

Table 1 shows the benefit of this technique. This table contains the number of rule heads on which unification would be attempted during execution of the different programs, followed by the number of successful inferences required of each. This number of inferences is constant for a given run, and the optimal situation occurs when the number of unifications attempted equals the number of inferences.

Table 1. Benefits of CAM Filtering

	Monkey and Bananas	Master Mind	Press
Attempts, No filter	3250	352	1703
Attempts, with CAM	1008	117	1367
Successful Inferences	829	104	998
Inferences/Attempt – no filter	25.5%	29.5%	58.6%
Inferences/Attempt – filtering	82.2%	88.9%	73.0%

Unification

Our CAM implementation assumes that each stack frame contains all the variable–value bindings made during its particular goal–head unification. During attempted unification new bindings are built, old ones referred to, and values compared. CAM helps all three.

New bindings from the unification of a variable with a value are built as multiword entries on the top of the stack, and contain the "name" of the variable (augmented by a stack depth to handle recursive rules) and its new value.

Dereferencing a variable to see if it has a value simply consists of a comparison against the names of all entries below the top of the stack in CAM. If no match exists, the variable has no bindings.

Unification is also simplified with CAM, and can be built into the dereferencing. When one argument is a ground term and another is a variable, at most two CAM cycles are needed. The first constructs a compare value consisting of the variable's name and the value it should have. A "match" means that the variable does in fact have that value as a binding. On an unsuccessful result, a second CAM cycle can determine if it is because the variable has a different value, or no binding at all. This is in contrast to the complex "tree" of comparisons needed conventionally.

More complex unification tests are equally helped by using variations of Figure 2, particularly if one or the other has embedded variables in it.

Overall Control

Another significant saving is in stack management, especially during a *backtrack* when the top frame and all its bindings must be removed. In conventional Prolog implementations, this requires support from yet another stack (the *trail*) built one entry at a time to record the address of variables that are bound to values. Then during an actual backtrack, a sequential "unwinding" through this trail is needed to reset the variables on it to an unbound state. With a CAM and the above stack algorithm, the trail is not needed, and the backtrack itself becomes a simple constant time operation to delete the frame.

Overall Performance Analysis

Table 2 gives some basic statistics on the Monkey and Bananas problem using a heavily instrumented Prolog interpreter built at IBM. The "Attempts using Indexing" row refers to a compiler technique where compile time information about the arguments of a goal predicts in advance the set of clauses which might be applicable. For reference, Table 3 gives a list of CAM operations performed for a slightly different form of the same problem, but with the above CAM algorithms integrated in.

Of particular interest here is the total number of CAM operations required, 235,618. For this version of the problem, this corresponds to 284 CAM operations per successful inference. With the CAM design of Figure 1, many of these can be overlapped with each other, meaning that less than 284 CAM machine cycles are needed per inference. This does not include machine cycles for the processor running the interpreter and controlling the CAM.

In comparison, an 80386 microprocessor running various compiled versions of this benchmark required anywhere from 340 to 2000 machine cycles per inference. The larger number came with versions performing asserts and retracts; the lower one came when there were no such knowledge base modifications, and the problem state was passed around as a dynamic data structure.

Table 2. Basic Execution Characteristics of Monkey and Bananas in Prolog

Attempted Unifications	3066	Maximum Stack Depth	429
Successes (Inferences)	688	Average Stack Depth	218
Attempts using Indexing	688	Backtracks	471
Bindings Per Inference	1.09	Bindings Unwound at Backtrack	475

Table 3. CAM Operation Counts for Monkey and Bananas

Operation	Count	Operation	Count	Operation	Count
Match First	29981	Read CAM	35383	Select Next	9175
Match Next	27200	Write CAM	4930	Shift RR Up	29602
DC Match	49774	Parallel Write	179	Shift RR Down	14946
Test RR	34426	RR op RR	22	Total CAM Ops	235618

From these statistics, and those of Table 1, we observe:

· Clause filtering using CAMs is a powerful idea, but one that can be matched by compiler technology when all clauses can be precompiled and indices to several lists of compiler-selected rules built. When there is a lot of *asserts* and *retracts*, however, compilers fail while the CAM technique will continue to work.

· Using CAM for the main Prolog stack significantly simplifies backtracking, but may require an unaffordable number of CAM words to handle the deepest inference chains. In fact, Monkey and Bananas is relatively benign in this regard; other problems we have looked at have maximum stack depths 10–100x deeper.

· Argument unification is also simplified with CAMs, particularly for complex objects. However, as with filtering, if the clauses are fully available at compile time, optimization techniques can often come close by building optimized comparison code for each argument (cf. statistics for the *WAM* model – Dobry *et al* 1985).

SUPPORT FOR PRODUCTION SYSTEMS

In contrast to Prolog, the programmer's view of a *production rule system* such as for OPS consists of separating the *rules* from the data *facts* (individually called *Working Memory Elements (WMEs)* and collectively called the *Working Memory* (WM)), and cycling thru a process that modifies these facts. Each WME typically has several fields that can have values. For example, in the Monkey and Bananas problem, a WME representing an object would have a field (its *class*) identifying its type, and others giving its name, where it is, how heavy it is, what it is on, what it contains, and what object can be a key to open it.

In the three step cycle the first *Match* step tries all combinations of WMEs against the *if* parts of all rules, and collects those combinations (called *instantiations*) that satisfy at least one rule into a *conflict set*. The *Conflict Resolution* step then selects one particular instantiation from the above set. This is passed to the *Act* step where the *then* part of that rule is executed. Usually this action adds, modifies, or deletes some WME, which then causes the whole cycle to repeat. An empty conflict set terminates the program.

Filtering

Generating all possible combinations of WMEs for potential matching against all possible rules is computationally unpleasant. One approach to reducing this is to take each WME as it is added to the system, test it against just those rules expecting WMEs of that class, and identify the successes. As with Prolog, this is called *filtering*. After filtering, the OPS system scans thru WM to identify WMEs that satisfy the other tests in these rules.

Again CAM is an excellent candidate for speeding up this process. For simplicity we assume two CAMs, one to hold WM, and the other an encoded form of the individual tests needed by the rules. One particularly interesting form of encoding creates a CAM entry for

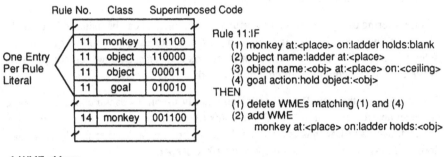

Figure 4. CAM–based Filtering

each literal in the *if* part of each rule. This entry contains the name of the rule, the class of the WME to be tested, and a field consisting of the superposition of a "k out of b" bit encoding of the values expected of different fields of WMEs of that class (see Figure 4).

Now when a new WME is created, all the values of its fields are encoded and OR–ed together. This is matched against the encoded rule CAM, with the result being a list of rules for which the WME may satisfy at least one test.

The second half of the match step would then take every rule identified by the first step, and look in the WME CAM for satisfying WME combinations. This is a multistep operation bounded in time by the number of literals in each rules, and the number of potentially matching WMEs. Unlike conventional approaches, this is independent of the total number of WMEs in WM. To demonstrate, in Figure 4, the *at* field from the monkey WME, is compared against the *at* field of all WMEs of class *object* whose *on* field equal *ceiling*. If there are no matches, rule 11 is not applicable. If there are matches we take them one at a time, and look into the CAM for WMEs of class *goal*, *action hold*, and *object* matching the value in the object's *name* field. Again no matches would cause that possibility to be discarded. Finally, another match cycle would look for objects of *name ladder* at the same *place* as the monkey.

Simulations are planned to determine the exact number of compare cycles needed, but some initial probability calculations have yielded the following estimates of CAM space and speedup over the brute force approach:

$$Rule\text{--}CAM\text{--}size = 48 \text{ Words} = C(\log_2 P + \log_2 L + B) \bmod 32$$

$$Speedup = 124 = \left(\frac{R}{2L}\right) * \left(\frac{D^2}{RM}\right)^R$$

where:

P = 20 = Number of Production Rules
C = 48 = Number of Condition Elements (Total)
D = 24 = Number of Distinct Attribute/Value Pairs
R = 61/48 = Average Number of Pairs/Condition Element
L = 3 = Number of of WME Classes
M = 3 = Number of Pairs Per Make
B = Number of Bits in Code Word (approx. 4)

Rete Nets

Current production systems clearly do not have CAMs, but achieve somewhat equivalent performance reductions by replacing the permutation process by one that compiles the *if* parts of all the rules into a *Rete net* (Figure 5). This dataflow–like net accepts *tokens* consisting of a WME and an indication of whether that WME is being added to or deleted from WM. Internally the net is constructed from two types of elements. *Alpha nodes* test one or more fields of a token and pass the token on only if the test is passed. *Beta nodes* have queues of tokens on each of two input, and upon arrival of a token representing a new WME at one input tests some field in that token against all the other tokens in the queue for the other input. Any token that passes the test is paired with the new arrival, and output for further testing. The original arriving token is also added to its appropriate queue.

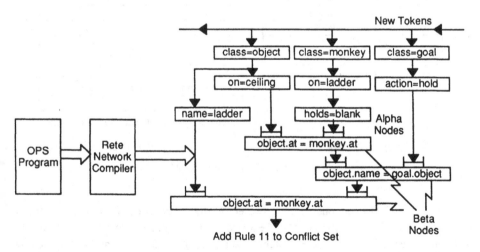

Figure 5. Rete Network Compilation

Tokens marked as deletions are handled similarly, except that instead of being added to a queue, their matching elements in the other queue are deleted. There is one output from the Rete net for each possible rule, and it delivers the same set of instantiations that would have appeared from the brute–force match cycle, but with a lot less work. Basically the beta nodes save internally all intermediate comparisons so that upon arrival of a new token only those pattern matches involving that token need be made. As an example, a somewhat optimized Rete net for the Monkey and Bananas problem contains 36 alpha nodes and 19 beta nodes.

In conventional implementations the beta nodes, especially handling deletes, consume most of the processing cycles. For example, in detailed simulations of several benchmarks, Gupta and Forgy (1983) found that the average token arriving at a beta node was compared against approximately 20 to 25 items on the other input, and that the average token triggered something approaching 35 beta node tests. In conventional computers this might correspond to in excess of 700 comparison instructions.

Again CAMs offer some unique capabilities here. Figure 6 diagrams one approach. One CAM (the alpha CAM) holds one entry for each string of pure alpha nodes in the Rete net. This entry contains the *class* of the WME passing through the alpha nodes, the expected values for the different fields being tested, and don't cares for the other field values. In addition, each entry also contains the name of the input queue of some beta node that these alpha nodes terminate in. Comparing a new token against this CAM thus in one operation identifies all beta nodes affected by it.

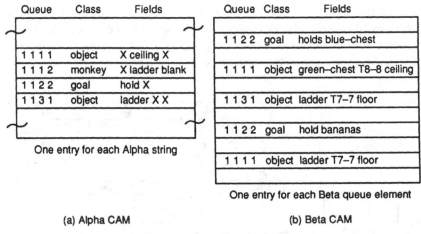

One entry for each Alpha string

One entry for each Beta queue element

(a) Alpha CAM   (b) Beta CAM

Figure 6. Rete Nets in CAM

The second CAM contains all the queue entries for all beta nodes. When a new token passes the alpha CAM, copies of it can be added to the designated queues in this CAM. In addition, for each queue, the other matching beta node queue number can be identified by table lookup, and a single CAM operation can thus match the appropriate fields of all WMEs in that queue against the beta node designated field in the new WME. The indicated matches can be pulled out one by one, paired with the new WME, and sent thru the alpha or beta CAMs as required by the Rete net. If the above statistics are correct, reductions on the order of 25 to one are possible for beta nodes, and a reduction to almost constant time for the alpha nodes.

An even more impressive time reduction occurs with the handling of tokens marked as *deletes*. Rather than passing them thru the Rete net as is done conventionally, we can in one CAM operation identify all queue elements that contain that WME, and mark them as now free. No alpha or beta search is needed. What used to account for almost half the processing time is now almost instantaneous.

Although we have not considered here all the variations of beta and alpha nodes that are needed in practice, the above analysis indicates that use of CAMs might speed up Rete based systems by perhaps a factor of 50 over conventional computers. Continuing work is under way to quantify this for a selected set of benchmarks, including the monkey and bananas problem.

Conflict Resolution

The Conflict Resolution step takes the set of instantiations identified by the Match step and selects one for execution. Current systems employ heuristics that attempt to prevent combinations from triggering more than once before all its *then* actions have been fully processed, and to make the process more sensitive to certain rule firing orders (such as the more recent, oldest, priority by rule,...).

A CAM can benefit both processes. Assume each instantiation is built out of a set of CAM words as discussed above. Clearly the second process often becomes essentially a constant time operation because it represents some sort of parallel search over the instantiations – an ideal CAM operation. Likewise the first can be greatly simplified by including in each set of CAM words representing an instantiation an extra field initially set to 0 when the instantiation is first added, and then set to 1 when the instantiation first fires. The selection process will ignore all entries with 1's set, while the process of entering new instantiations will not. A match of a new instantiation with one already in the set, even if its new field is set, will prevent that instantiation from being added, thus preventing looping. Upon completion of the rule, the CAM words containing the instantiation in the conflict set can be cleared out as discussed above, again in constant time.

SUMMARY

In this paper we have shown the usefulness of CAMs in supporting rule–based AI languages. In particular we have shown the importance of properly designed row logic to support sophisticated pattern matching operations.

The techniques for supporting Prolog use such features extensively. While modern Prolog compiler technology can get a good part of the same gain without CAM by optimized code generation, these gains are not possible when there is a lot of dynamic program modification. CAM, on the other hand, continues to handle such cases nicely.

The only problem with CAM and Prolog is Prolog's need for larger stack spaces in the near future than may be economically viable with CAM. This implies that some sort of hybrid CAM/RAM scheme may be the most cost effective approach.

In contrast CAMs and modern OPS compiler technology seem to complement each other perfectly. After compilation the major operations are still pattern matches over relatively short lists – excellent matches to current CAM densities. Large speedups seem attainable, and will be verified in the coming months.

Finally, a topic of growing interest is implementing truly parallel Prolog–like languages. Such languages introduce new problems, particularly with multiple copies of bindings for the same variables, that today's compiler technology has great difficulty with, but which appear to be an excellent match to a CAM. This and the more detailed studies of CAMs and OPS are the focus of our future work.

ACKNOWLEDGEMENTS

The University portion of this work has been supported by the New York State Center for Computer Applications and Software Engineering at Syracuse University, the Systems Integration Division of IBM Corporation, and Coherent Research Inc. The National Science Foundation provided educational access to the MOSIS chip fabrication brokerage.

REFERENCES

Colomb, R.M., "A Clause Indexing System for Prolog Based on Superimposed Coding," in *The Australian Computer Journal*, vol. 18–1, 1986.

Dobry, T.P., Despain, A.M., and Patt, Y.N., "Performance Studies of a Prolog Machine Architecture," in *Proc. 12th Symp. on Computer Architecture*, pp.180–190, 1985.

Gupta, A. and Forgy, C., Measurements on Production Systems, CMU–CS–83–167, Carnegie Mellon University, Dept. of Computer Science, 1983.

Oldfield, J.V., Stormon, C.D. and Brule M.R., "The Application of VLSI Content–Addressable Memory to the Acceleration of Logic Programming Systems" *Proc. CompEuro*, pp. 27–30, 1987.

Sodini, C.G. and Wade, J.P., "Dynamic Cross–coupled Bitline Content Addressable Memory Cell for High Density Arrays," *IEEE J. Solid State Circuits*, vol. 22, pp. 119–121, 1987.

Sohi, G.S., Davidson, E.S., and Patel, J.H., "An Efficient Lisp–Execution Architecture with a New Representation for List Structures," in *Proc. 12th Annual Symp. Computer Architecture*, pp. 91–98, 1985.

Stormon, C.D., Brule, M.R., Oldfield, J.V. and Ribeiro, J.C.D.F., "An Architecture Based on Content–Addressable Memory for the Rapid Execution of Prolog", *Proc. of the Fifth International Conf. and Symposium on Logic Programming*, pp. 1448–1473, Aug. 1988.

4.2 UNIFY WITH ACTIVE MEMORY

Yan Ng, Raymond Glover and Chew–Lye Chng

INTRODUCTION

Since the christening of the 5th Generation Computer Project in Tokyo, 1981, the research activity in PROLOG machines has gradually gathered momentum. Important developments include the Warren Machine (Tick and Warren 1984, Dobry *et al* 1985) that focuses primarily on compiler technology, the MIMD approach which uses multiple processors (Codish and Shapiro 1986, Conery and Kibler 1981) and the SIMD approach that is based on a content Addressable Memory (CAM) (Robinson 1986, Oldfield *et al* 1987). They all seem to provide significant improvements in performance and this poses the problem of which is the best architecture for a PROLOG machine.

This paper is intended to investigate the mapping between some aspects of PROLOG constructs and their hardware support. As unification is one of the most frequently performed tasks in the execution of PROLOG program, it has been chosen as the focus of this investigation. The number of unify operations and the proportion of its execution time in relation to the whole of the search process have been studied. Simulations show that about 50% of the PROLOG activity deals with unification (Civera *et al* 1987). Another simulation study (Woo 1985) indicated that unification function accounts for 55% - 70% of the query processing time. These studies were conducted independently, although their findings were subject to particular implementations of PROLOG, nevertheless, in both cases, it was concluded that unification would be the prime candidate for acceleration.

THE IMAGE MACHINE

As the unification function relies mainly on pattern matching, hardware acceleration for the matching operation has always been a focus of attention for speed improvement. This could be achieved by the use of dedicated hardware register stacks with the necessary logic wired to speed up the matching operations. An order of magnitude in speed improvement had been reported with this approach (Nakazaki *et al* 1985). Nevertheless, we believe a serious mismatch exists between the image machine (unification) and the host machine (the hardware accelerator), variables must be encoded in a way to get different treatment in the unification process. Furthermore, in order to cater for lists, built-in predicates, constants and variables, the hardware units need to be fine-tuned with microprogramming.

However, study of the operational behaviour of unification function has revealed the following pattern illustrated by the example given in Table 1.

Table 1 The 'Facts' and 'Rules' of Unification

GOAL CLAUSE :   parent (john, mary).

\--

FACT CLAUSES: parent (mary, jane).
parent (fred, lucy).
parent (bill, mary).
::   ::   ::
father (john, mary).
mother (anne, mike).
::   ::   ::

\--

RULE CLAUSES: parent (X , Y):-
father (X , Y).
parent (X , Y):-
mother (X , Y).
::   ::   ::

\--

Unification is an attempt to match a goal clause against a set of rules and facts, and in that process more subgoals may be generated. This process will go on until the proposition can be proved, or until such time that all options have been exhausted. Therefore, looking at the architectural mapping of this operation, we propose a table processor, capable of handling predicates, lists and structured terms with variables and constants; able to report on the outcome of all matching operations at every stage of the unification process; and able to support variable bindings within the memory structure.

Translating these into the specification of the image machine shown in Figure 1, requires a SIMD operation for broadcasting the goal clause to the entire database of rules and facts, a response register for reporting the outcomes of all matching clauses for further investigation and facility to support variable binding.

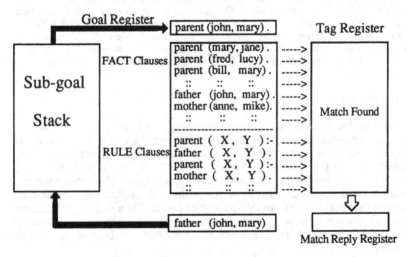

Figure 1 The Image Machine for Unification

THE HOST MACHINE

Based on the above requirements, a host machine using associative memory is proposed to perform searching on a SIMD basis (Figure 2). The goal clause is loaded into the Input Data Register (IDR) which can simultaneously compare with every bit of every word-row within the Associative Memory Array (AMA). The outcomes of all matching words are then tagged on either the Tag Register TR1 or TR2, and the overall result is reported in the Match Reply Register (MRR) to be used as a condition for any branching instruction. However, in order to select binding clauses, mechanisms to isolate and activate word-rows are supported by the Word Select Unit (WSU), with the introduction of tag manipulating operations between the SEARCH and READ/WRITE operations. Furthermore, in order to accommodate clauses of variable length, such as lists and structured terms, a word length of 40 bits has been chosen as a compromise , which on the one hand is sufficient to store a predicate or an atom, yet on the other hand avoids redundancy by allowing the cascading of an unlimited number of word-rows for lists and structured terms.

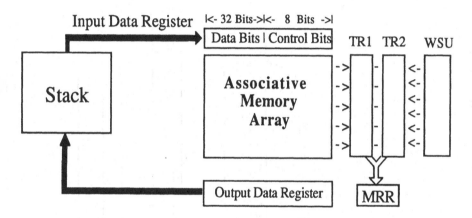

Figure 2 The Host Machine for Unification

THE DATA ORGANISATION

Consider a typical PROLOG predicate such as *parent (john, mary)*, it is becoming apparent that its data representation in textual form would be very inefficient, and some form of binary representation such as that defined below will be advantageous for data compaction.

> *<atom> ::= <term>,<identity><status><control bits>*
> where *<term>* is mapped onto an internal binary code representing an atom,
>> *<identity> ::= V / C* {as being variable or constant}
>> *<status> ::= HD / L / S / P / . / , / :-*
> This field is used to identify atoms as being predicate, list, structured term, predefined predicate, or various delimiters.
>> *<control bits> ::= <B><B><B><B>*
>> *<B> ::= 0 / 1 / X*

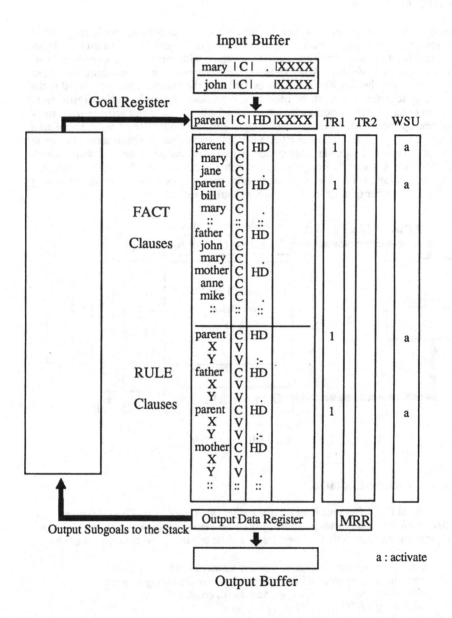

Figure 3 The Data Organization of Atoms within the AMA

THE ARCHITECTURE

The CAM based approach for the support of pattern matching has been reported by Oldfield (1986, 1987) and Robinson (1986) with a tremendous speed up in the acceleration of unification for certain specific tasks such as variable bindings, heap processing, and clause filtering. However, pattern matching only constitutes part of the unification problem, there is also the problem of variable binding, or using Warren's dictum, the "logical variable":

unification = pattern matching + logical variable (Warren 1977)

Oldfield *et al* demonstrate several schemes using CAM support for variable binding, aimed at 1000-KLIP performance with a CAM-based accelerator and a RISC based host. However, where multiple matching of rule heads occurs, the authors believe a closer mapping to architecture could be achieved with minor enhancements to the CAM array. A double-layer CAM array is, therefore, being proposed for the handling of rule clauses where variable binding could be accommodated within the memory. Elsewhere in this volume, Kogge, Oldfield, Brule and Stormon discuss alternative approaches, including clause filtering.

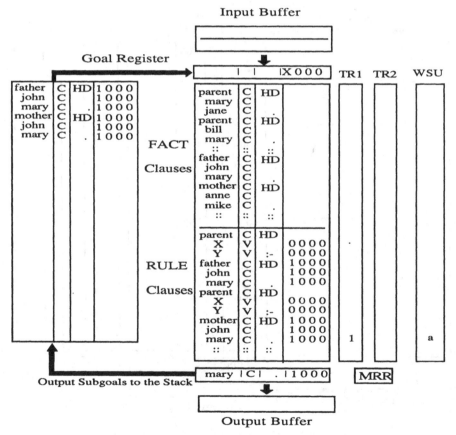

Figure 4 Variable Binding within AMA

Consider the database shown in Figure 3, since the predicate *parent(john, mary)* does not exist among the fact clauses in the database, the logical proposition is for the binding of *john* with *X* and *mary* with *Y*. Two subgoals will be generated from the main goal of *parent(john, mary)*. Hence, the ideal solution is to create a mechanism, whereby variables of interest will be pushed into the background temporarily, and where binding could then take place. After the binding, the state of the database is shown in Figure 4.

In reality, the implementation of a double-layer CAM array will incur a fair amount of redundancy. However, not all word-rows of rule clauses need to be provided with the double layer CAMs, only those word-rows containing variables need the provision of the background layer to be used later in the event of binding. Hence, a set of instructions coupled with hardware support have been designed to serve this purpose. Background layers are created and reserved during the loading of the database (facts and rules) as shown in Figure 5. Instructions are provided for the database to switch from one layer to the other during the unification process. In other words, before binding, the rule clauses are shown in Figure 3, and after binding, the rule clauses should look like those in Figure 4.

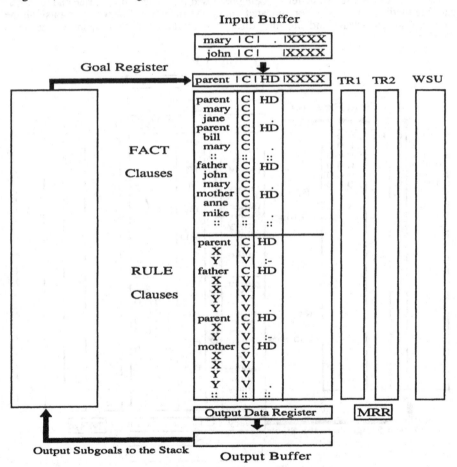

Figure 5 Word-Row Provision for Binding

At computational organization level, the associative Processing Instructions (APIs) are organized into an Associative Computational Cycle, corresponding to major clock beats at the microinstruction level.

```
I<--- Phase 1 -->I<------- Phase 2 --------->I<--------- Phase 3 --------->I<-- Phase 4 --->I
1st SEARCH  -> 2nd SEARCH/CLEAR -> TAG MANIPULATION -> READ/WRITE
```

Every computational cycle normally consists of two instructions, namely <API1> and <API234>, with <API1> starting the search operation, and followed by either one of the <API234> instructions depending on the outcome of the Phase one search. In Phase three, a set of tag manipulation operations have been designed to select or activate matched word-rows and/or their adjacent rows, these form the basis for information retrieval and data transformation.

1. API1: SEARCH Operation

   The API1 instructions, first of all, initialise TR1, WSU and MRR, before comparing the data held in IDR (or ODR) with Contents of AMA. All matching word-rows are tagged in TR1 and subsequently set the MRR if at least one match have been found.

2. API2: SEARCH Operation

   The API2 instructions perform a similar operation as that of API1, except the matching responds are now tagged in TR2, instead of TR1. The purpose of the API2 search is for marking blocks of word-rows for group activation.

3. API3: Tag Manipulations for the activation of neighbouring word-rows.

   The function of API3 operations is to provide the linkage from the known search key to the unknown information. The connection is established by the propagation of tag or tags from matched rows to other adjacent word-rows.

4. API4: WRITE/READ Operations.

   This is the final phase of the Associative Computational Cycle in which a READ or a single/multiple WRITE operation will take place on activated word-rows.

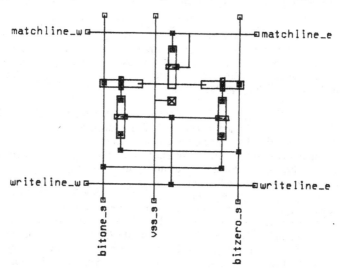

Figure 6 The Symbolic Layout of Wade/Sodini Cells

VLSI IMPLEMENTATION OF CAM FOR VARIABLE BINDING

This section describes the potential of the concurrent variable binding algorithm for VLSI implementation. Pattern matching can be accomplished by straightforward CAM implementation which can take the form of a five MOS transistor cell such as that proposed by Wade and Sodini (1987) in Figure 6.

The use of CAM for storing expressions requires a re-interpretation of how items are normally matched during a parallel search of the CAM. Variable entries in expressions must match any search key and then be bound subsequently to the value of the search key Table 2.

Table 2 Requirement for Matching Logical Variables

IDR	AMA	Results
$\overline{Var}$	$\overline{Var}$	Depends on the content of both IDR and AMA
$\overline{Var}$	Var	Match
Var	$\overline{Var}$	Match
Var	Var	Match

This can be accomplished by associating a flag bit with variable entries in the CAM which forces a match for all flagged rows. The flag column can be disabled by a global control line to enable searching for a particular variable. Two implementation strategies are possible: the flag bit could be contained as part of the CAM row which would require a two pass search first to activate all variables which are part of a matching clause or clauses, or logic could be provided associated with the flag column to enable a match without a pre-activating search. Design of a test chip is proceeding with the latter implementation.

In order to maximise the concurrency of variable bindings it is necessary to replace the variable contents of all matching expressions in parallel with their bound value when the match is made rather than when they are transferred to the bindings stack. This requires temporary storage for each variable expression so that it can be restored should unification of the particular sub-goal fail. There are three strategies that have been investigated to accomplish temporary storage:

1) Conventional CAM is used with a spare row allocated to each variable. This requires bit serial, row parallel copying of the variable which is considered to be too great an overhead with the CAM row of 40 bits.
2) Space could be made for variables by shifting them in adjacent unused word rows using the shifting CAM described in Ng and Glover (1987). This CAM cell requires 19 transistors and is rejected because of its size in this application.
3) A compromise shifting CAM cell using 10 transistors is possible here because it is only necessary to preserve information in one row of a pair of rows. The CAM rows are connected up and down by single-phase pass transistor elements. The Wade-Sodini cell is used as the basis for the CAM and the data is sensed in a storage element by pre-charging the data line columns and then enabling an appropriate up or down control line for each row to copy the data onto the CAM row above or below.

It may be noticed that in all strategies both bound and unbound expressions are stored in CAM which can be used to halve the number of times variables have to be copied and restored. In order for the normal tag register propagations to operate transparently around temporary storage rows it is necessary for tag register elements to be controllably connected to adjacent and next-but-one adjacent bits. A control register contains marked temporary storage rows and is used to control looping over unused tag register bits.

Arrays of CAM cells have been implemented using the 2μ CMOS process provided by ES2 (European Silicon Structures) in order to evaluate their potential. The layout of the single-phase shifting CAM has a cell size of 68μ by 41μ and its layout is shown in Figure 7.

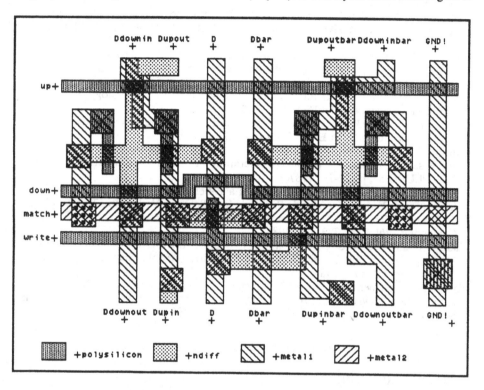

Figure 7 Layout of a Single-Phase Shifting CAM Cell

THE CONCLUSION

From the research and investigations conducted, the authors are convinced that a CAM based table processor can form a basis for a hardware accelerator for the unification function. However, in order to minimize the overhead for variable binding, a set of tag manipulation instructions have been proposed for the implementation of variable binding within the Associative Memory, which is unique among other CAM based accelerators.

Preliminary studies of the chip architecture and layout suggest that chips containing 100 word rows of 40 bit words of modified CAM are feasible in 2μ technology.

REFERENCES

Civera, P., Piccinini, G. and Zamboni, M. "VLSI Architecture for Direct Prolog Language Interpretation", COMPEURO 87, pp.168-172, 1987.

Codish, M. and Shapiro, E. "Compiling OR-parallelism into AND-parallelism", 3rd International Conference on Logic Programming, Springer-Verlag, pp.283-297, 1986.

Conery, J. S. and D.F. Kibler, D. F."parallel Interpretation of Logic Programs", Proc. ACM Conf. on Functional Programming languages and Computer Architecture, pp.163-170, 1981.

Dobry,T.P., Despain, A.M. and Patt, Y.H., "Performance Studies of a Prolog Machine Architecture", ACM SIGARCH Newsletter, Vol. 13, Issue 3, pp.180-190, June 1985.

Nakazaki, R. et al, "Design of a High-Speed Prolog Machine (HPM)", 12th Annual International Symposium on Computer Architecture, pp. 191-197, 1985.

Ng, Y. H. and Glover, R. J. "The Basic Memory Support for Functional Languages", Comp Euro 87, First International Conference on Computer Technology, Systems and Applications, Hamburg, 11-15th May, 1987.

Oldfield. J. V., "Logic Programs and an Experimental Architecture for their Execution", IEE Processings, Vol. 133, Pt E, No.3, pp.163-167, May 1986.

Oldfield, J. V., Stormon, C. D. and Brule, M. "The Application of VLSI Content-addressable Memories to the Acceleration of Logic Programming Systems", COMPEURO 87, pp.27-30, 1987.

Robinson, I. "A Prolog Processor Based on a Pattern Maching Memory device", 3rd International Conference on Logic Programming, Springer-Verlag, pp.172-179, 1986.

Tick, E. and Warren,D.H.D. "Towards a Pipelined Prolog Processor", New Generation Computing, Vol. 2, pp.323-345, 1984.

Wade, J. P. and Sodini, C. G. "Dynamic Cross-Coupled Bit-Line Content Addressable Memory Cell for High-Density Arrays", IEEE Journal of Solid-State Circuits, Vol. SC-22, No.1, pp.119-121, 1987.

Warren, D. "Implementating Prolog" Tech. report 39, Edinburge University, May 1977.

Woo, N. S. "A Hardware Unification Unit: Design and Analysis", 12th Annual International Symposium on Computer Architecture, pp. 198-205, 1985.

ACKNOWLEDGMENTS

This work was supported in part by Science and Engineering Research Council fellowship B/ITF/65. The authors wish to acknowledge the helpful discussions with John Oldfield regarding this paper.

4.3 THE PATTERN ADDRESSABLE MEMORY: HARDWARE FOR ASSOCIATIVE PROCESSING

Ian Robinson

INTRODUCTION

Operations involving pattern matching are ubiquitous amongst artificial intelligence applications. Many such fall into the class of searching rule bases for applicable entries via some form of pattern match. Such operations are so fundamental that they have become the basic execution mechanism in production systems and logic programming. However performing the actual pattern match over a rule database is computationally intensive. To offset this various techniques have been developed to move this complexity to compile-time. Examples include the Rete algorithm (Forgy 1982), compilation of Prolog clauses into sequential WAM code (Warren 1977), the building of hash indices and superimposed codeword techniques (Wise and Powers 1984).

By their very nature, such techniques are best suited to systems involving static rule-bases with rigidly defined and limited forms of access. As systems are built that require more flexible and non-deterministic access to the rule-base (by, for example, a complex 'reasoning system') and/or give rise to a more dynamic rule-base (eg. hypothesis generation and test, 'genetic algorithms' (Davis 1987)) then such techniques become beset by overhead problems which seriously hamper performance; problems that increase with the size of the database. As artificial intelligence applications become more complex and interactive these traits are bound to come to the fore, forcing the complexity issue back into the run-time system, and the realm of special purpose hardware.

THE PAM

The pattern-addressable memory (or PAM) chip (Robinson 1986, Anderson *et al.* 1987) belongs to the class of *associative processors* (Thurber and Wahl 1975, Yau and Fung 1977, Foster 1976). It combines SIMD parallelism with the 'smart memory' approach to high processing bandwidth: the processor logic is replicated amongst the storage cells on the chip. The PAM's storage is content addressable via a hard-wired pattern match algorithm. There is no provision, or requirement, for directly addressing physical locations.

The PAM system is made up of an array of these chips and a controller. Regardless of the number of chips the PAM system is designed to behave as a single linear array of storage slots, written as a stack. Data, whether read, write or match query, is carried on a global data bus to all slots. The controller simply mediates instruction and data traffic between the PAM chips and the host system.

Symbols and expressions

Given the application domain outlined above the PAM is oriented towards symbolic, rather than numeric, processing. A symbol in this context occupies one word of memory. A variety of symbol types are supported, each identified by type tags. Strings of symbols form *expressions* which are essentially similar to LISP s-expressions. These are entered as written, i.e. with sub-structure expanded out in place. Expressions and sub-structure are delimited by parentheses. To distinguish the top level, expressions are separated by a 'spacer' symbol.

Within these delimiters the contents of an expression are made up of *constants*, represented by their names ('a','b' and 'c' for example), and *variables*. Variables play the traditional database role of 'don't care's' and can be either named or anonymous. They come in two types described below. Both the stored expressions and the input *query* expression have this same syntax. The ability to handle stored don't care's is something not often provided in other associative processing systems, be they hardware or software. It is important however as it allows the system to manipulate incomplete information. In general the syntax can be seen to be very 'Prolog-like'. This is intentional as Prolog is able to subsume most database query languages.

Pattern matching and the match engines

With regard to the pattern matching algorithm a parenthesis only matches another parenthesis with the same 'handedness', and constants only match other constants with the same name. A *one-place variable*, represented '?X', '?Y', etc. or anonymously as '?', matches any constant, variable or sub-expression. A *cdr variable*, on the other hand, will match the *cdr* of a list. More generally it matches any number, including zero, of consecutive symbols within an expression.

An important feature of expressions is that they are of no fixed length and thus cannot be easily wedded to the fields of traditional CAM memory organisations. To deal with this expressions are stored in physically consecutive words in the PAM and the query is input sequentially, i.e. symbol by symbol. Each word has associated with it a 'match engine' that together compute and track the progress of matches through the stored expressions (a scheme similar to Lee (1963)). The progress of matches through the stored expressions are marked by 'match markers' that are passed on from symbol to symbol as they match, or destroyed if they don't. An example of this process is shown in Figure 1.

Each slot contains storage for a symbol, its associated type tag bits, a status tag bit (the match marker), and a match engine. The match engine in turn is comprised of a comparator, between the name parts of the stored word and the input, and a small finite state machine to control the progress of the match marker. In keeping

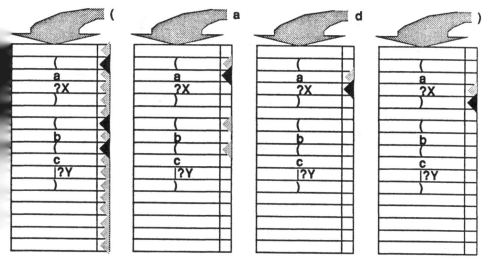

Figure 1 Progress of match markers given the query '(a d)'
(Previous marker positions are shown shaded)

with the requirements of the pattern matching algorithm the FSM is also used to implement matching other than that of the constant-constant variety (which is what the comparators are there for). To this end it takes as inputs, along with the comparators output and the match marker from the previous word, the type tags of the stored word and the query.

Apart from matching there are other functions that also require their own status bits. For example most data i/o is mediated by the 'read marker' and the 'write marker', and expressions are deleted via 'delete markers'. Control of these markers is exercised by the FSM via an instruction word broadcast to all the match engines, one of its states, of course, being *match*.

Jumping and skipping

The 'jump-wire' is a mechanism connecting all the match engines and allows markers to be moved within and between expressions without the slot-by-slot limitation imposed by the FSM match engine. An example of its application occurs during pattern matching when a stored sub-expression may be required to match against an input one-place variable. We require that the match marker present at the beginning of the sub-structure be 'jumped' to its end. Moreover this must occur in parallel with potentially other matching expressions containing similar sub-structure, and yet others with a one-place variable in that position.

Similarly, in common with other CAM architectures, there is the problem of match resolution: given that there may be several responders to a particular query, how do

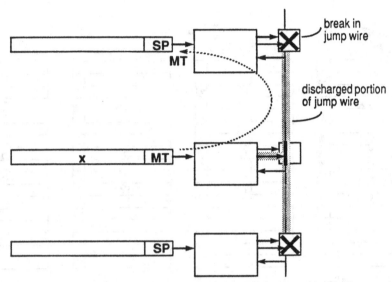

Figure 2 Moving a match marker to the head of an expression using the jump-wire
(The large boxes represent the FSM's)

you pick one at a time for output? In the PAM the completion of the match query leaves the responders with match markers at their ends. First these markers are moved back, in parallel, to the heads of their respective expressions. Subsequently the read marker is initialised at the top of the memory array and then jumped to the first match marker. After being used to mediate output of that expression it can then be jumped to the next, and so on.

The jump-wire is controlled by the match engine's FSM. Essentially it consists of a single wire running the length of the memory array that is precharged on every cycle and then selectively split up and discharged. In the match resolution example (see Figure 2) the match markers are moved by first 'breaking' the jump-wire at the spacer symbols between expressions. A discharge is then made to occur wherever there is a match marker. This causes only the sections of the jump-wire corresponding to matching expressions to be discharged. Each match engine that is associated with a spacer and sees a discharged jump-wire below itself then creates a new match marker.

The sub-structure match is similar except that markers are created at *all* closing delimiters in the remainder of the expression, as the PAM has no way of distinguishing the matching one. This can be tolerated as the remainder of the pattern match will weed out any 'false' markers in all but a few pathological cases. These latter can be guarded against by paying closer attention to parenthesis balancing and the arity of expressions.

In matching there is also the inverse case to consider: that of a stored one-place variable matching an input sub-structure. In this case the match marker must be held at the location of the variable until the sub-structure has been skipped. This is handled by an additional state bit called the 'skip marker'. A match marker enabling a match on a one-place variable is stashed away as a skip marker if the input symbol is an opening delimiter. A match marker is subsequently created whenever a closing delimiter is input.

The case of the cdr variable is similar except that match markers are set on every subsequent *symbol*, not just closing delimiters.

'No match' detection

In addition to the jump-wire a simple wire-OR is strung between the match engines. This is used to detect the absence of match markers. Using this a match can be interrupted if no stored expressions continue to match the query, effectively 'answering before the question is complete'. In many applications a large percentage of queries will be met by a *no_match* response, and this mechanism provides about the fastest possible way of delivering it.

Multiplexing

The complexity of the match engines due to this range of supported functions is, unfortunately, reflected in their size. With one match engine physically attached to each storage slot, as implied heretofore, overall memory density would be very poor: a drawback shared by most previous CAM designs. To improve on this a word-serial scheme (similar in concept to Crofut and Sottile (1966)) is used, effectively multiplexing each match engine over a number of storage slots. In the PAM the match engines cycle, in lockstep, through their respective storage words, updating their status, for each query symbol entered. An artifact of this scheme is that storage is divided into *pages* as shown in Figure 3. When a page is selected it is as though every slot on that page has its own match engine. Circuitry is provided to allow expressions and markers to pass from one page to the next.

In general then matching times increase by a factor equal to the number of pages, 'P'. However using the no-match circuitry, now on a per page basis, advantage can be taken of clustering like expressions onto one or a few consecutive pages. The eventual absence of match markers on the *other* pages is used to 'prune' them out of the page cycle. Performance is therefore increased as the bulk of the matching, and subsequent operations, will ignore the bulk of the pages.

THE CHIP

The physical layout of the PAM chip consists of a number of replicated elements, called *blocks*, and some peripheral control and i/o circuitry (see the chip micrograph in Figure 4). Each block corresponds to a horizontal slice through Figure 3, as indicated by the shading. As such it contains a match engine and a block of memory P words deep addressed by an external page counter. The word selected by the current page number is the one input to the comparators and the FSM. A schematic of a block is given in Figure 5.

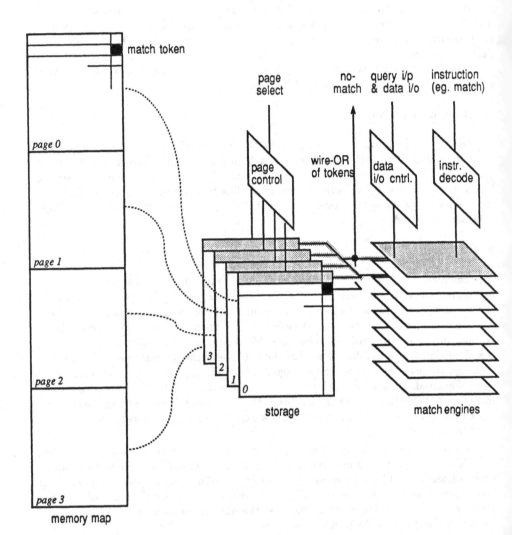

Figure 3 Chip schematic showing the page multiplexing scheme
(Only four pages and eight match engines are shown for clarity)

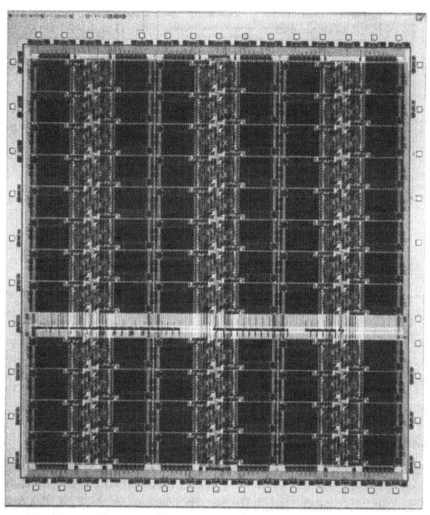

Figure 4 Micrograph of the PAM20 chip.

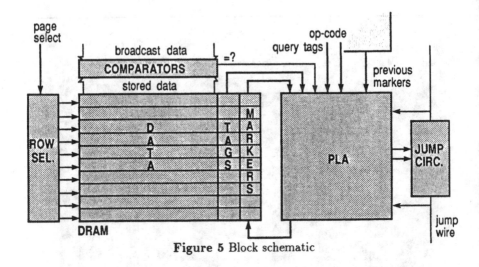

Figure 5 Block schematic

Memory and comparators

The memory is simple three-transistor DRAM. A cycle has four phases corresponding to precharge, read, precharge and write. This incorporates the read-modify-write of the markers via the FSM, and the refresh of the stored data by means outlined below.

Previous CAM designs have used a dual rail scheme for the bit and query broadcast lines enabling a bit-wise 'don't care' capability used in a number of numerical algorithms (Kohonen 1980, Foster 1976). This is not required in the PAM and a more compact single rail scheme is employed using complementary transistors in the comparator to achieve the required function (Figure 6).

Refresh and i/o

Refresh is handled via a dummy word of storage in each block. On each page selection the addressed words are also written to these stores, and then written back to their sources along with the new state of the markers generated by the FSM. In this way a whole page is refreshed in one cycle. Provided all pages are accessed reasonably frequently no special refresh routine need be run during the course of normal processing.

This extra word of storage is modified to allow its contents to be written from and read onto the global data bus, thus facilitating data i/o. Reading from and writing to unique locations in the PAM are controlled via the read and write markers, there being only one of each in the entire PAM. Typically the write marker marks the next empty slot in the PAM's memory array, i.e. the location to which the next input word should be written. The read marker is moved as desired in the operations described earlier (and is also responsible for enabling its current host chip onto the PAM system's data bus). Each marker automatically advances by one slot on its respective read or write instruction.

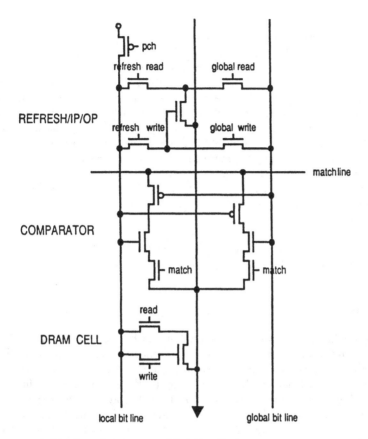

Figure 6 A bit-slice showing one DRAM cell and its connections to comparator, refresh and data i/o circuitry

Modifying stored data

In common with other associative processor designs the PAM has a *multi-write* capability. This allows the input symbol to be written to all locations marked by a match marker. The multi-write function opens the door to actual processing within the PAM subsequent to pattern matching. It can be used to tag all matching expressions or to update in parallel all instances of a particular variable, for instance. Another variation of the write command performs a write to a location just read via the read marker. This allows individual entries to be selectively updated, e.g. incremented, or a bunch of matching expressions to be given unique markers, for example. Both these modify functions can be extended to whole sub-expressions.

Deleting

Deleting an expression consists of three steps. First it must be selected. A special form of matching is used to do this called *exact matching*. As the name implies the matching algorithm used is that every stored symbol must match every input symbol constant for constant, variable for variable etc.. Without this capability it would be difficult to mark, say, a more general form of a rule without also marking more particular forms.

Secondly the deleted expression must be marked as such using delete markers. The jump-wire is employed to set delete markers throughout the expression that matched the previous step. The presence of delete markers prevent an expression from taking place in any further matches; so, to all intents and purposes, the expression is no longer there. It does however still occupy space. This space can be reclaimed at any later time using the third process: garbage collection. The garbage collection routine uses a sequence of internal reads and writes to compact out all words marked with a delete marker.

The delete marker, in its role of marking empty space, is also used to mark those memory locations not yet written to. The entire PAM can be emptied by a single instruction that sets the delete marker in every word.

CONCLUSIONS

The prototype PAM chip shown in the micrograph of Figure 4 has a capacity of 1152 20-bit words (which, including tags and markers, amounts to approximately 32k bits) distributed over 16 pages. It is fabricated in $2\mu m$ double-metal P-well CMOS, occupies a die size of $7853\mu m$ by $8694\mu m$ and has a cycle time (for the four phases) of 200ns. Actual pattern matching performance is very dependent on the nature of the expressions due to the page-pruning scheme. Simulations with a number of scenarios suggest a typical match and readout time of approximately $6.5\mu s$ for queries with an average length of ten words, significantly faster than traditional hashing schemes. These experiments also show that processing time is relatively insensitive to query length but far more dependent on the percentage of no-matches, followed by how many cycles it takes to 'prune' the matching down to the relevant pages. Note that performance is not strongly affected by memory size (i.e. the number of PAM chips used) due to the SIMD nature of most operations.

The flexibility of the PAM's instruction set makes it a potentially high performance addition to many symbolic processing systems where such concurrency can be exploited. Many parameters of the PAM design, e.g. word length, multiplexing ratio and even the nature of the match engine FSM, are quite flexible. This allows architectures of the same general type to be tailored to other applications where a combination of 'smart memory' functionality, high performance and good memory density are required.

Acknowledgements

This paper describes work done by the author whilst at the Schlumberger Palo Alto Research Center. The author would also like to take this opportunity to thank Al

Davis, John Conery, Shimon Cohen, Gary Lindstrom, and the many and varied other collaborators on the late FAIM-1 project for their help in developing the ideas behind the PAM.

References

Anderson, J. M., Coates, W. S., Davis, A. L., Hon, R. W., Robinson, I. N., Robison, S. V. and Stevens, K. S., "The Architecture of FAIM-1," *Computer,* vol. 20, pp. 55–65, 1987.

Crofut, W. A. and Sottile, M. R., "Design Techniques of a Delay-line Content-addressed Memory," *IEEE Transactions on Computers,* vol. 15, pp. 529–534, 1966.

Davis, L. ed., *Genetic Algorithms and Simulated Annealing.* Los Altos, Ca.: Morgan Kaufmann, 1987.

Forgy, C. L., "RETE: A Fast Algorithm for the Many Pattern/Many Object Pattern Match Problem," *Artificial Intelligence,* vol. 19, pp. 17–37, 1982.

Foster, C. C., *Content Addressable Parallel Processors.* New York: Van Nostrand Reinhold, 1976.

Kohonen, T., *Content Addressable Memories.* Berlin: Springer-Verlag, 1980.

Lee, C. Y., "A Content Addressable Distributed Logic Memory with Applications to Information Retrieval," *Proceedings of the IEEE,* vol. 6, pp. 924–932, 1963.

Robinson, I. N., "A Prolog Processor based on a Pattern Matching Memory Device," in *Proceedings of the Third International Conference on Logic Programming,* E. Shapiro (ed), London, UK.: Springer-Verlag, pp. 172–179, 1986.

Thurber, K. J. and Wahl, L. D., "Associative and Parallel Processors," *Computing Surveys,* vol. 7, pp. 4, 1975.

Warren, D. H. D., *Implementing Prolog.* Tech. report 39: Edinburgh University, 1977.

Wise, M. J. and Powers, D. M. W., "Indexing Prolog Clauses via Superimposed Code Words and Field Encoded Words," in *Proc. Int. Symp. Logic Programming,* pp. 203–210, 1984.

Yau, S. S. and Fung, H. S., "Associative Processor Architecture – a Survey," *Computing Surveys,* vol. 9, pp. 1, 1977.

Chapter 5

KNOWLEDGE BASED SYSTEMS

Knowledge representation and manipulation is required in a number of artificial intelligence algorithms. Knowledge representation has been the subject of intense study in the artificial intelligence community for some time (IEEE 1986). The goal in knowledge representation is to allow AI programs to behave as if they know something about the problems they solve. Schemes for knowledge representation include logic programming, semantic networks, procedural interpretation, production systems and frames. The use of knowledge representations can be found in automated deduction systems, inference machines, expert systems and knowledge bases.

Processing large knowledge representations requires an enormous computational power that may not be delivered by a conventional uni–processor architecture at a reasonable speed. Therefore, special purpose architectures are needed to manipulate such knowledge representations. A major requirement for knowledge based systems is to perform pattern matching over a large number of objects and the relationships between these objects.

ARCHITECTURES FOR KBS

Special–purpose hardware has been developed for knowledge based system: DADO (Stolfo 1987) a rule–based machine and The Connection Machine (Hillis 1985) and NETL (Fahlman 1979) semantic network architectures.

In this chapter two architectures for KBS are described. Lavington *et al* §5.1 present a relational algebraic processor (RAP) which implements operations such as *intersection* and *join*. The RAP achieves high performance by exploiting the parallelism in relational operations in a modularly extensible manner. The RAP is shown to satisfy many of the computational requirements of large knowledge–based systems, including the implementation in hardware of graph manipulations such as transitive closure and shortest path.

Delgado–Frias and Moore §5.2 describe a wafer scale architecture for semantic networks. This defect–tolerant multiprocessor architecture handles knowledge bases that are represented in a semantic network form. The defect–tolerance approach is based on a combination of hardware redundancy and robust algorithms run in the architecture. The application studied in this paper is *scene labelling*; the matching process is also explained.

References

Fahlman, S. E., *NETL: A System for Representing and Using Real-World Knowledge.* Cambridge, MA: The MIT Press, 1979.

Hillis, W. D., *The Connection Machine.* Cambridge, MA: The MIT Press, 1985.

IEEE, "Special Issue on Knowledge Representation," *Proceedings of the IEEE,* vol. 74, no. 10, pp. 1299–1450, October 1986.

Stolfo, S. J., "Initial Performance of DADO2 Prototype," *Computer,* vol. 20, no. 1, pp. 75–83, January, 1987.

5.1   A HIGH PERFORMANCE RELATIONAL ALGEBRAIC PROCESSOR FOR LARGE KNOWLEDGE BASES

Simon Lavington, Jerome Robinson and Kai–Yau Mok

INTRODUCTION

The principal requirement for knowledge-based systems is the ability to perform pattern-directed search over large numbers of variously-sized objects.   The precise interpretation of the word 'object' depends upon the knowledge representation formalism being used. Examples of objects range from simple numbers or variables, through tuples or logic clauses, to quite large objects such as relations, semantic nets or procedures.   Objects are conceptually held in some form of associative (i.e. content-addressable) store.   Pattern-directed search of this store is a process of recognition or selection, which in general results in a *set* of objects being discovered to match the search criteria.   The output from a search operation may be called the *responder set*.   AI-related processing consists, implicitly or explicitly, of further operations on responder sets.   Amongst these operations, set intersection and transitive closure occur so frequently that they have been proposed, along with pattern-directed search, as prime candidates for hardware support for AI (Fahlman et al 1983).   Furthermore, set intersection is one of a family of relational algebraic operations which is central to all database-like computation.   Finally, transitive closure is just one of a family of graph-manipulation operations (other examples are shortest-path and activation-propagation)   which find a use throughout Computer Science.

There is clearly a need for special-purpose hardware which stores and manipulates sets and graphs.   Existing hardware which offers at least part of the desired functionality divides into four categories: (a) disc-based back-end database machines such as the ICL CAFS (Maller 1980);   (b) RAM-based relational co-processors such as the Ferranti Relational Processor (Dixon 1987); (c) special-purpose devices such as the Generic Associative Array Processor chips (McGregor 1987); (d) SIMD arrays of single-bit PEs and RAM such as the Connection Machine (Hillis, 1985) or the AMT DAP (Parkinson et al 1988).   In the context of large knowledge bases, category (a) suffers from limited flexibility and speed,   category (b) from limited flexibility and capacity, and category (c) from limited functionality and capacity. Category (d), being in essence a collection of general-purpose processors, offers rich functionality but the cost per bit is excessive if the whole of a large knowledge base is to be contained within the SIMD array.   There is thus scope for a new approach.

There are several architectural issues that have to be addressed, in order to ensure that novel hardware matches the requirements of large knowledge-based systems. The issues may be divided into two (overlapping) groups: memory-dominated and processing-dominated. For the former group the choice of appropriate low-level unit of associative storage is the most important issue. This is naturally related to knowledge representation formalisms, and to the practical necessity to construct a total memory system from a hierarchy of (semiconductor and disc) technologies. For the processing-dominated issues, the central theme is the maintenance of a smooth physical and conceptual interface between memory (e.g. the knowledge base) and processor (e.g. the inference engine) throughout the computation cycle. On the physical side, this becomes a matter of minimising the movement of information between sub-units and maximising the data-bandwidth when movement is unavoidable. On the conceptual side it is desirable that the computational model takes advantage of the natural set- and graph-related attributes of associative storage.

Devising suitable computational models is currently the subject of much research. An associative store is essentially a parallel device, and operations such as set intersection contain obvious natural parallelism. It is thus desirable to move away from sequential computational models such as linear input resolution. Parallel computational models for functional languages, e.g. based on graph reduction, are showing promise; however the set is not normally a primitive in these languages. There is some evidence that computational models for logic languages, such as Bibel's Connection Method (Lavington *et al* 1988), give better scope for parallelism than resolution; in addition, the Connection Method can usefully employ relational algebraic operations.

The above architectural issues are discussed further in (Lavington 1988). The practical outcome is a system known as the Intelligent File Store (IFS), which provides a framework within which to implement hardware support for large knowledge bases. In the next section we introduce those aspects of the IFS which establish the context for the design of a special-purpose sub-unit known as the Relational Algebraic Processor (RAP). The objective of the RAP is to provide rapid processing of responder sets obtained from the IFS's associative store. The command repertoire and target performance are then given. The detailed RAP hardware design is presented in the section which follows. Finally, the overall performance is discussed.

THE IFS ARCHITECTURAL FRAMEWORK

The IFS contains several sub-units, of which the principal one is an Associative Predicate Store (APS). The APS consists of an associatively-accessed disc and a semiconductor associative cache, unified by a memory management scheme known as semantic caching (Lavington *et al* 1987). The APS cache is modular in design, with a total capacity extending to several tens of Megabytes. The prototype IFS contains a 4 Mbyte cache. The search time, which is independent of capacity, depends upon the number of wild cards in the interrogand and is typically 250 microseconds plus one microsecond for every responder.

The format of each line, or tuple, stored within the APS is fixed. It may be configured by software at start-up time to suit the needs of particular knowledge representation formalisms (Lavington 1988). Briefly, the layout of each APS entry is as follows:

$$<L><T_1><T_2>...<T_m><M_1><M_2>...<M_n>$$

where:

> $<L>$ is an optional label (Gödelisation, etc)
> $<T_1><T_2>...$ are terms in a wff
> $<M_1><M_2>...$ are optional modifiers for the $<L>$ and $<T>$ fields.

The above represents the allowable APS tuple options as presented at a software IFS Procedural Interface. The APS hardware is actually semantics-free, within the practical constraints of the present implementation which limits the total number of $<L>,<T>$ and $<M>$ fields to 15 and the total length of a tuple to 63 bytes.

The compact length of each tuple is made possible by the use of Internal Identifiers (IDs) to represent primitive objects. IDs are of fixed length, configurable at start-up time within the range 2 - 8 bytes. The top four bits of each ID denote its type. Three types are system-defined, namely:

> string — (used for external lexemes)
> numeric — (used for integers)
> label — (used for Gödel numbers, etc.)

The remaining 13 types are information-model dependent, and may be used to represent abstract nodes, existential variables, etc. The IFS provides hardware assistance with allocation, deallocation and mapping into and out of the ID space. For type string, an associative store called the Lexical Token Converter (LTC) handles strings of up to 120 characters in length, with fuzzy searching provided for the mapping from string to ID. In the IFS prototype the LTC has a capacity of 2 Mbytes, with search times in the range 10 - 450 microseconds (independent of capacity).

The above simple structures have enabled bulk associative stores to be built for the IFS. A consequence of the IFS architecture is that functionality is distributed amongst several sub-units. In the IFS prototype these sub-units communicate via a VME bus, which also contains a Sun 3/75 CPU with 4 Mbyte of RAM. The IFS Procedural Interface runs on this Sun, which also allows the IFS to be a network-accessible resource.

As far as relational algebraic operations are concerned, the APS provides equivalence selection of responder sets which may be passed to the RAP for further processing - (see below). Fuzzy selection and non-numeric range selection are performed via the LTC. Numeric range-selection, aggregate functions, and all types of sorting will be carried out in a stream processing unit presently being designed.

IMPLEMENTING RELATIONAL ALGEBRAIC OPERATIONS IN THE IFS

The principal operations of the Relational Algebraic Processor (RAP) are intersection, difference, union, join and unique. (Unique is the removal of duplicate tuples from a set). Derivative graph operations are discussed later.

A number of operations have common features, making possible their implementation on the same special-purpose hardware. Intersection, Difference, Union, and Unique are all based on the Membership operation. Membership is easy to implement in hardware because it requires only Yes/No output. Operations on two sets require interaction between each element of one set with all elements of the other set. If we arrange that one set is distributed amongst a number of Search Modules, each one equipped with RAM and comparator, then a SIMD design is made possible which can exploit the natural parallelism inherent in relational algebraic operations.

The Membership operation has the advantage that no data output is needed from Search Modules. The system operates as a *filter* for input tuples. The Membership result that allows a tuple through the filter depends on the relational operation: Intersection allows it through if it *is* a member of the set in the Search Modules; Difference, Union, and Unique allow it through if it is *not* a member.

Join is similar to the Membership-based operations. It is an extension of Intersection, but differs in ways that complicate the hardware: i) it uses Partial Matching; ii) it requires output from the stored set. Only part of the input tuple is broadcast to Search Modules for matching. The output required if a match is found can include the associated data from the input tuple or from the matched tuples in the Search Modules. A data output path must be provided from Search Modules, and because more than one tuple in the stored set may match the broadcast searchkey there is a conflict for output. An Output Controller is therefore needed to supervise data output from Search Modules, and to create the required format output tuple by concatenating specified parts of the input tuple with the output from each Search Module.

RAP hardware provides equi-join on a key which may be any collection of fields chosen from the tuples in each set. Tuple size and field organisation may be different in the two sets. Tuples to be joined can be any length. Tuples loaded into the prototype Search Modules are each limited to 512 bytes maximum, but tuple size in the other set is unrestricted. Tuples resulting from the join can be any collection of fields chosen from the pair of tuples whose keys matched, and the fields can be assembled into any specified order to create the new tuple.

In the first instance all IFS sets originate as APS responders, and are therefore collections of <L> and/or <T> and/or <M> fields as defined previously. The data interface between the RAP and the rest of the IFS is by a central semiconductor memory buffer. Sets are transferred in and out of the RAP by DMA via the VME bus to this central buffer. The RAP requires storage for one of the sets. In a 2-set operation the first set flows into the RAP and is stored. Then the second set flows through the RAP, generating the result set as output. The memory buffer (Search Buffer) holding the first set is actually the array of Search Modules. Each Module is a block of memory plus a Search Engine, so it operates as a bit-parallel word-sequential content-addressable memory (CAM). The APS and RAP both use this form of CAM which has been called quasi-associative (Hahne, 1985). Associative retrieval is obtained by each Module searching its memory for the specified (input) data item. There is no restriction on the size of set that can be handled by the RAP. If a set is too large to load completely into the Search Buffer, the operation becomes a sequence of subset operations, organised by a RAP controller (see below). The result is that overall operation time increases when a small RAP (one with few Search Modules) is required to work on a large set.

RAP HARDWARE STRUCTURE

Tuples passing through the RAP follow a three-stage pipeline: Input and Output modules sandwich the parallel array of Search Modules. The arrangement is coordinated by a RAP Controller, as shown in Figure 1. The various sections of Figure 1 are now described more fully.

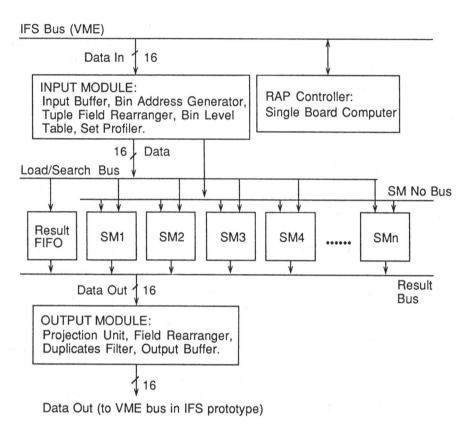

Figure 1 RAP Architecture showing main data paths.

RAP Controller

Different sub-units in the IFS system operate concurrently to implement a group of operations. The RAP Controller (which is actually a Single Board Computer, SBC) is able to accept and queue requests for operations while the RAP hardware is working, so there is no hardware setup delay between operations. The SBC initialises registers before each hardware run. It communicates with the hardware through an Instruction Queue: a FIFO for opcodes and parameters. The SBC can write to the instruction queue at any time, even when the hardware is working on a previous operation. The hardware will download from the queue as soon as it completes its present operation. The SBC also handles any exceptions that occur during hardware operation, and provides self test facilities and assistance with hardware fault location.

The Input Module

Sets arrive in the RAP a (16-bit) word at a time, via the IFS prototype's VME bus. They pass through the Input Module which contains a tuple field rearranger and a hash address generator. The hash address selects a hash bin in the Search Array, and all Search Modules (SMs) in the array search that bin in their memory for the input tuple. The input tuple is broadcast on the Load/Search Bus - (see Figure 1) - to all Search Modules, which perform their search in lock-step.

The Input Module performs a number of different tasks. It provides input synchronisation for the SM array, which deals with tuples rather than words. It can rearrange the sequence of tuple words on arrival, to produce a tuple whose fields are in a different order: ABCD becomes CBDA, for example. This facility is useful in sequential comparison when fields to be compared in the two sets are in different positions and/or order.

Tuples are, as it were, self-indexing. The start address used for the search is derived from the input tuple via hashing. A hardware unit in the Input Module generates this hash address.

The Input Module maintains a table of Bin Level information. During loading each hashing bin is filled in a way that ensures tuples spread evenly (within the bin) across all Search Modules (whole tuples in each SM). The table contains two items for each bin:

1) The number of the SM to load with the next tuple
2) The start address to use when loading.

The start address is an offset from the hash bin address. The offset remains constant until all SMs have a single tuple in the bin, which then contains one *tuple layer* and the offset entry is increased by the constant amount which is the number of words in tuples. The offset then remains fixed until the next tuple layer is complete. SM number is incremented by one every time a tuple loads into the bin. In addition to its use during the LOAD phase, the Bin Level Table is used during SEARCH. It serves two functions:

1) Allowing termination at end of data, not end of bin
2) Generating the Incomplete Tuple Layer signal (ITL).

The offset entry in the Bin Level Table indicates the number of words to be shifted out of the bin to the comparators. However, the top layer of tuples may be incomplete (not all SMs contain tuples in that layer). When that final layer is being compared, only SMs whose number is less than the Table entry should be active. The ITL signal allows the Result Logic in each SM to deactivate itself if its SM number is >= the value on the SM No Bus, which was read from the Bin Level Table.

If the two entries in the Table are zero for a particular bin, then that bin is empty and need not be searched. The Input Module is thus able to filter out tuples where no search is required - they are not passed on to the Search Array stage. This idea is carried further in the Input Module, which also maintains a Set Profiler: a bit map whose address is derived from selected tuple bits. If the map bit is zero during search then the current input tuple is not in the Search Buffer. If the bit is set then a search is performed to discover whether the tuple that set it was the same as the current input.

The data interface between Input Module and Load/Search Bus behaves as a FIFO with retransmit facility for each tuple. The ability to retransmit is needed when searching bins containing more than one tuple layer, and during Unique.

The Search Array

The Search Array is a one-dimensional array of Search Modules, with two controllers to synchronise the SMs as a SIMD system. SMs all receive the same control signals.

The Search Module structure is shown in Figure 2. It is based on Video RAM (VRAM), which is a special-purpose memory combining 64K x 4-bit DRAM and 256 x 4-bit SRAM on the same chip. The Search Buffer is therefore a modular memory whose capacity can be incremented in 64K x 16-Bit modules. Increasing the Search Buffer size by adding Search Modules allows larger sets to be accommodated, and increases speed of operation - because each module is a separate search engine. This is an example of extensible parallelism.

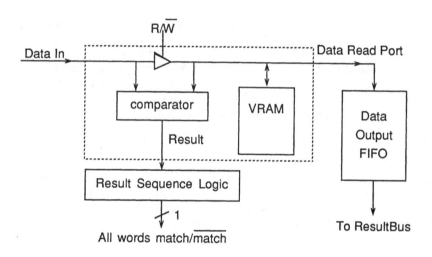

Figure 2   Search Module, showing proposed VLSI module structure.

The SRAM holds a relatively small number of words, namely, a row of DRAM words. The block of 16-bit words that transfers in a single step between the two memories is a DRAM row (256 words in current devices). The chip is designed for continuous data output, without pausing during transfers, so all 64K words are accessible, and can be shifted out at full speed for continuous sequential searching. This feature allows the full searchrate to be maintained continuously - with no delay between successive inputs. The hash bin can be changed without missing a comparison cycle.

DRAM requires refresh, but this can be done in the background while a row is shifting out of SRAM to a comparator, so there is no pause for refresh. Full speed searching can proceed continuously, and the number of words compared per second is the shift rate of the SRAM - currently guaranteed up to 25 MHz, which is 25 million 16-bit word comparisons per second.

The FIFO in Figure 2 is used in Join, Transitive Closure and Graph Path operations to avoid the need to stop searching while output is in progress.

The Output Module

The Output Module in Figure 1 contains a Projection Unit which can remove fields from output tuples. It can also change the order of fields if required.

Projection may generate duplicate tuples. A Set Profiler is provided as a (coarse) filter to identify duplicates. It is a 1 MBit RAM addressed by any 20 bits selected from the tuple. This is sufficient to distinguish between tuples whose variation is known to be limited to 20, or fewer, specified bits. To remove duplicates from sets whose tuples contain greater variation, the RAP's "Unique" operation is applied to the output set.

Scope for a VLSI-based Design

That portion of the Search Module structure shown within the dotted border in Figure 2 is a suitable candidate for VLSI. The plan would be to modify an existing component which is readily available, and widely used, as a commercial VLSI product: Video RAM (VRAM). An extension to the VLSI structure of VRAMs could produce a single-chip Search Module, useful for rapid sequential searching and general associative access, and cascadable for longer wordlength. VRAMs are dual-port RAM, but SMs use only the Serial Port. Redundant pins and I/O chip area therefore exist in Search Module VRAMs.

PERFORMANCE

The RAP is presented to system software via a relational algebraic interface which has the notion of sets and relations as built-in primitive types. This interface is in turn supported by a group of RAP-specific commands which are being added to the existing IFS low-level Procedural Interface (Lavington 1988), written in C.

Search Rate

The maximum number of 16-bit word comparisons per second is limited by VRAM access and cycle times. Maximum access time for the HM53461 VRAMs used in the RAP is quoted as 40 ns, and we find typical values around 30 ns. A searchrate of 20MHz (50 ns cycle time) was therefore easily obtained with a SM built from off- the-shelf components. Faster VRAMs are anticipated as manufacturers try to meet demands for increased graphics resolution and colour - which entail rapid data access from VRAMs.

20 MHz searchrate is 20 M comparisons per second. Our current wordlength is 16 bits, so the comparison rate in each SM is 40 MBytes per second. If the wordlength was increased to 32 bits, the searchrate per SM would increase to 80 MBytes per second. An array of ten 32 Bit SMs would provide a searchrate of 800 MBytes per sec,and an array of 50 SMs, 4 GBytes per sec.

The number of SMs in the array is extensible. Increasing the number increases the size of set that can be stored in the search buffer, and improves the average searchrate. A minimum array size of about 16 SMs is envisaged, because of the complexity of the associated control logic compared with the simplicity of SMs. We hope by using surface-mount components to fit 8 SMs onto each double-height Eurocard. (The wire-wrapped prototype is of course less dense than this). The maximum number of SMs in the proto-type Search Array is the number that fit into a 19 inch subrack. This could naturally be extended in the future.

We have carried out an evaluation of the RAP when performing a Join operation, choosing synthetic data so that a comparison can readily be made with other implemen-tations. Two relations, A and B, each of 1024 tuples, were joined on a common field. Values in this common field were generated randomly with the following constraints: Values in A range from 0 to 63; those in B, from 60 to 123. There are thus four common values. Each relation has about 16 occurrences of each value. This scheme therefore produces about 4*16*16 tuples as output.

Each tuple of A and B consists of: a 16-bit unique key; the value field described above; and other fields sufficient to produce a total tuple length of 64 bytes. When the relations A and B were joined in the Ferranti Relational Processor (Dixon 1987), the elapsed time was 90 milliseconds, assuming relations A, B already exist within the RP; a further 300 msec were taken to form the resulting relation. The same Join operation performed in the RAP will take 3.2 msec, ignoring the VME Bus delay in loading the relations and removing output because the RAP is operating faster than the VME bus bandwidth. In general, the RAP will give higher performance for high-cardinality rela-tions. Note also that, using IFS IDs, the tuple size of 64 bytes could represent up to 32 fields. RAP performance improves with fewer fields.

One final point concerning performance: hashing loses the original order of tuples in relations. It is possible to retrieve the original order by tagging tuples during LOAD with a one word count field, and then using a counter in the join operation to generate the input sequence that brings the tuples out in their original order. 64K 64-word tuples can be retrieved in order in about 250 msec by this method. 32K 32-word tuples take about a tenth of a second. This constitutes a Sort algorithm, but is only suitable for short wordlength key fields because the second set in the Join is all possible values of the key, in order - (64K for a 16-bit key)

USE OF THE RAP FOR GRAPH OPERATIONS

Structures can be represented as sets of tuples and processed by relational operations. A graph may be a set of node-arc-node triples. Each node is one or more fields. This representation of network links (graph edges) allows arbitrarily complex structures to be stored as sets of fixed size tuples, easily processed by the existing RAP hardware.

Many Knowledge Base applications use transitive closure rules (such as supervisor, descendant, linked path, component). These rules are specified by a linear recursive rule cluster, which contains a single linear recursive rule and one or more non-recursive (exit) rules:

$$a(X,Y) :- b(X,Y).$$
$$a(X,Y) :- c(X,Z), a(Z,Y).$$

Han *et al* (1988) show that other, more complicated, recursive rules (which are not Transitive Closure rules) can also be evaluated by transitive closure, after preliminary compilation. They also compare the efficiency of algorithms for evaluating transitive closure and found the delta-Wavefront algorithm gave the best performance:

```
WAVE := a;        (* WAVE is unary relation *)
CLOSURE := a;        (* CLOSURE is a unary relation *)

While WAVE ≠ nil do
Begin
        WAVE := WAVE ∇ B;  (* B is the base relation *)
        WAVE := WAVE - CLOSURE;  (* Set Difference *)
        CLOSURE := WAVE ∪ CLOSURE  (* UNION *)
End.
```

Each iteration generates a new wave of nodes in the connection graph, by applying the ∇ operator. This "quarter Join" operator performs the operation:

$$\Pi (y) [A(w) \bowtie B(w,y)]$$

- where B is a base relation and A is a set of IDs to be matched with one of the two fields in B. The other field of B is output on match: match field w and output field y. The set of IDs output constitutes the new "Wavefront".

The two lines:

```
WAVE := WAVE - CLOSURE;  (* Set Difference *)
CLOSURE := WAVE ∪ CLOSURE  (* UNION *)
```

together constitute the RAP's "Unique" operation. The RAP therefore provides fast implementation of the delta-Wavefront algorithm.

Using similar compositions of basic RAP operations we can implement other graph manipulations. Examples are: "find shortest path from node X to node Y"; "is there a path from node X to node Y passing through nodes {A, B, ...}?". The use of the RAP for such tasks is the subject of a forthcoming paper.

CONCLUSION

When designing special-purpose hardware, there is a danger of being over-specific to the extent of producing a "solution looking for a problem". The Relational Algebraic Processor described in this paper has been designed to fit into an architectural framework, namely the IFS, which can be shown to support the functional requirements of large knowledge-based systems. A production IFS/1 has been installed at the Artificial Intelligence Applications Institute, Edinburgh, for independent evaluation.

In addition to providing the usual relational operations such as join and intersection, the RAP may be used for graph manipulations such as transitive closure and shortest path. Furthermore, the RAP provides high performance by exploiting the parallelism inherent in all these operations in a modularly extensible manner. Finally, the central part of the RAP's Search Module has been shown to be a prime candidate for fabrication as a VLSI component.

The RAP is the subject of UK Patent Application No. 88 070230.

ACKNOWLEDGMENTS

The RAP has been developed as part of Alvey funded projects GR/D/45468 -IKBS/129 and GR/E/05018 - IKBS/066. The work of Kai-Yau Mok is partially supported by the K. C. Wong Education Foundation and the Croucher Foundation. The authors take pleasure in acknowledging the many helpful discussions with members of the IFS team.

REFERENCES

Dixon, K.B., "The Ferranti DVME 785 Relational Processor". *Internal report 6902, Oct. 1987, Ferranti Computer Systems Ltd.*

Fahlman, S.E., Hinton, G.E. and Sejnowski, T.J., "Massively parallel architectures for AI: NETL, THISTLE, and Boltzmann machines". *Proc, Nat. Conference on Artificial Intelligence (AAAI-83)*, Washington, pp109-113, 1983.

Hahne, K., Pilgrim, P., Schnett, H., Schweppe, H. and Wolf, G., "Associative Processing in Standard and Deductive Databases". *4th International Workshop on Database Machines*, March 1985.

Han, J., Qadah, G. and Chaou. C., "The Processing and Evaluation of Transitive Closure Queries". *Proc. Int. Conference on Extending Database Technology*, Venice, March 1988. (Springer-Verlag: 303).

Hillis, W.D., "The Connection Machine". *MIT Press*, 1985.

Lavington, S.H., Standring, M., Jiang, Y.J., Wang C.J., and Waite, M.E., "Hardware memory management for large knowledge bases". *Proc. of PARLE, the Conference on Parallel Architectures and Languages Europe*, Eindhoven, June 1987, pages 226 to 241. (Published by Springer-Verlag as Lecture Notes in Computer Science, Nos. 258 & 259).

Lavington, S.H., "Technical Overview of the Intelligent File Store". *Knowledge-Based Systems, Vol. 1, No. 3*, pp 166-172, June 1988, .

Lavington, S.H., Jiang, Y.J. and Azmoodeh, M., "Towards an implementation of Bibel's Connection Method". To be presented at the 1988 UK IT Conference, Swansea, July 1988.

Maller, V.A.J., "Information retrieval using the Content Addressable File Store". *Proc. IFIP-80 Congress*, pages 187-192, published by North-Holland, 1980.

McGregor, D., McInnes, S. and Henning.M. "An architecture for associative processing of large knowledge bases". *Computer Journal, Vol.30, No.5*, pp 404-412, Oct. 1987 .

Parkinson, D., Hunt, D.J., MacQueen, K.S., "The AMT DAP 500". *Proc. 33rd IEEE Computer Society International Conference*, San Francisco, pp 196-199, Feb. 1988.

5.2 A WSI SEMANTIC NETWORK ARCHITECTURE

José Delgado–Frias and Will Moore

INTRODUCTION

A number of artificial intelligence applications needs reliable high performance computers that allow them to execute algorithms at reasonable speed. In recent years, new architectures have been developed in order to accomplish the performance required (Uhr 1987, Wah and Li 1986). Many AI algorithms use knowledge representation and manipulation which can be accomplished by means of semantic networks. A WSI architecture dedicated to semantic network processing may have an impact on the AI field since the parallelism in this application can be exploited by means of wafer–scale integration technology.

Wafer–scale integration (WSI) technology offers the potential for improving reliability and performance of large integrated circuit systems (McDonald et al 1984, Peltzer 1983). The reliability of any packaged integrated device is improved when the number of pins is reduced since mechanical connections between different packaging levels are the main failure sites in any chip. Speed can be improved by having short interconnection lines among the system's modules. However, WSI yield losses caused by gross and random defects are the major limitation on WSI technology success (Stapper et al 1983, Mangir 1984). Gross defects usually originate from manufacturing errors and cause a large portion of the wafer to lack functionality. Random or *spot* defects may originate from errors or defects in photolithography, process quality, or starting silicon material. Random defects are far more frequent than the gross defects (Peltzer 1983); therefore a wafer–scale architecture must be able to tolerate spot defects.

Quillian introduced the idea of *semantic networks* in 1968. Several proposals and changes have been made since; Brachman (1979) provides an overview of these proposals. The basic principle of a semantic network is simple (Winston 1984): each node of the interpreted graph represents a *concept* and the links between them represent *relationships* between concepts. Several architectures have been proposed for semantic network manipulation (Delgado–Frias and Moore 1988).

A semantic network multiprocessor architecture has a knowledge base (KB) stored in the array. In order to make inferences the facts or queries must be sent to the architecture in a semantic network form. The machine will match the incoming network with the KB; when a match is found variables are assigned with a value.

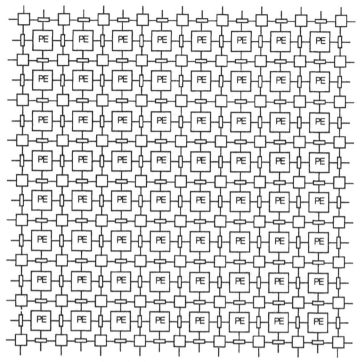

Figure 1 Semantic network array

SEMANTIC NETWORK ARCHITECTURE

A semantic network architecture that is designed to be implemented on wafer scale integration should be able to tolerate silicon defects as well as incorporate all the advantages of the technology.

The proposed WSI architecture consists in principle of a two–dimensional array of processing elements (PEs). This architecture, shown in Figure 1, operates in single instruction multiple data (SIMD) stream fashion; a central controller sends macro–instructions to all PEs. Each PE has a micro–controller which allows it to make some decisions, such as to pick up data on which the macro–instruction is to be executed and to search for communication paths in order to send data. Sending macro–instructions not only requires less global communication but also provides enough time for broadcasting the instructions. While PEs are executing the previous macro–instruction the new one is broadcasted. A novel communication scheme provides paths between a PE and its eight nearest neighbours as well as row and column interconnection. A PE stores and manipulates several nodes and links of the semantic network. The building blocks of the architecture are described below.

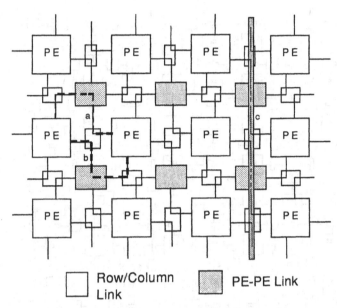

Figure 2 Interconnection scheme

Communication Scheme

A knowledge base (KB) in semantic networks is represented by nodes which have relationships specified by links. The nodes and links are computed by PEs of the semantic network architecture. Dedicated lines to all possible connections between nearest neighbours is not recommended in WSI technology since silicon defects affect some communication lines. A dynamic or soft communication scheme between PEs is required in order to deal with the faults and provide alternative communication paths. If the communication scheme circuitry has few transistors and small area the yield and reliability are expected to be high.

The communication scheme, shown in Figure 2, allows each PE to communicate to any of its eight nearest neighbours. The paths are bidirectional and provide two independent connections between two nearest neighbours; this is accomplished by programming the PE–PE links as shown in the Figure (doted lines between the PE in the centre of the array and the one on the middle of the left column). In order to avoid complex modules and large amount of hardware, no storage in the communication network is allowed. It is also possible to arrange the array into a row and column communication scheme by programming the row/column links. A column is shown in Figure 2, the links should be programmed as shown by the doted lines.

The PE–PE link module, shown in Figure 3, has only six transistors per communication line. When communication between a PE and any of its eight nearest neighbours occurs, a single transistor exists between PEs; this not only provides minimum delay but also high reliability and yield. The module has six control lines that can be programmed by any of the PEs next to it.

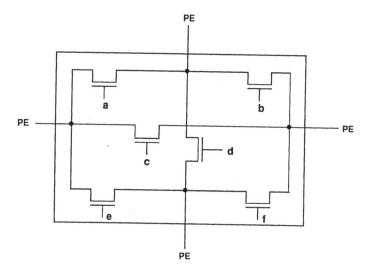

Figure 3 PE–PE link module

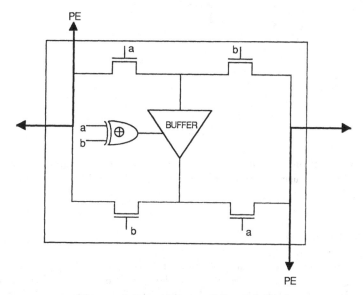

Figure 4 Row/Column link module

The Row/Column link module, shown in Figure 4, has a buffer in order to drive signals along a column or row. This buffer is set unactive when row/column communication is not needed or bidirectional communication is required. Signals are driven in an unidirectional manner at a given time, but the direction can be changed. Each PE is connected to two rows and two columns and has two paths for nearest neigh-

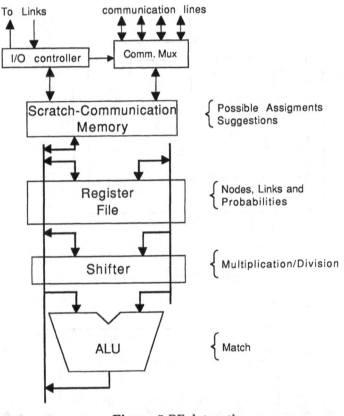

Figure 5 PE datapath

bour communication. Alternative paths are available in this communication scheme; these paths can be used when either a component failure blocks communication or a path is busy.

Processing element

The semantic network processing element (PE) is able to match nodes and links as well as a probabilistic measure of believe. The type of operations that a PE performs are not only symbolic but also include arithmetic and communication tasks. Thus, the PE must handle these operations in order to obtain higher performance.

The PE datapath, shown in Figure 5, performs communication and processing tasks independently; the former is mainly executed by the I/O controller and the latter by the PE's ALU. These two tasks are synchronized by writing and reading messages from the scratch and communication memory. This memory also allows storage of temporary results, such as possible assignments (when a node is matched the variable and its measure of certainty are stored), suggestions of possible match (once a node in

NODE

node link (NIL)	SYMBOL	Proba-bility	ADDRESS			ADDRESS			ADDRESS			Cont
			PE	REG	I/O	PE	REG	I/O	PE	REG	I/O	

LINK

node link (T)	SYMBOL	Proba-bility	ADDRESS			ADDRESS			ADDRESS			
			PE	REG	I/O	PE	REG	I/O	PE	REG	I/O	

BUFFER

node link (NIL)	SYMBOL <buffer>	ADDRESS			ADDRESS			ADDRESS			ADDRESS		
		PE	REG	I/O	PE	REG	I/O	PE	REG	I/O	PE	REG	I/O

Figure 6 Node, link and buffer formats

the neighbourhood is matched this may send suggestions) and matched links. In order to speed up multiplication and division a shifter is used. The shifter needs not be large, since the maximum integer number that is to be compute is 100; thus, an eight–bit shifter may be large enough to carry out the operations. Logical and arithmetic operations are executed in the ALU; matching is the most important instruction.

The register file is used for storing nodes, links and probabilities which all are part of the semantic network knowledge base. The nodes and links are matched to input nodes and links that are stored in the scratch and communication memory; while the probabilities are used to compute a measure of believe (MB) which is an evaluation of the match goodness. Figure 6 shows the format in which nodes and links are stored; forty–six bits are required for this format. The first field is a Boolean one that is F when a node is stored and T for a link. A symbol is stored in the second field whose size depends on the number of objects (or concepts) as well as the distance between symbols. The probability field in the link representation is an indication of its weight which represents the uniqueness of the link. In the node that probability gives the maximum value of the measure of believe the node; this feature allows to have nodes with low and high connectivity connected by means of a link (such a link may affects each node differently). Address fields give information about node–link connections as well as the direction of information. The PE address is relative in order to have simple decoding hardware and small address words.

Since some nodes may have many links there is a need for handling more than three link connections as provided by the node format. This can be accomplished in two ways: first, by using the CONT field which is last in the node format. When this field is set to T it indicates that the address list continues in the next register which has the buffer format. Second, if the node requires more than seven links then buffers are put in different PEs since having buffers in the same PE causes problems with connectivity. The fanout, that is the number of links that can be handled by a

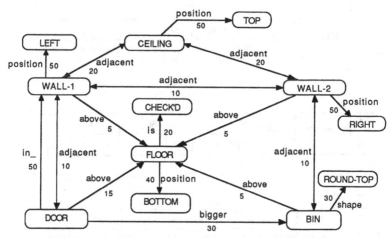

Figure 7 Semantic network of a scene labelling system

node, can be extended by the use of *buffers* which cause some delay. The relationship between buffer–delay D and maximum fanout F_{max} is: $F_{max} \approx 7 \times 4^D$.

Fault–tolerant strategy

A novel fault–avoidance approach has been developed to deal with silicon defects that affect part or entire PEs. The aim of this approach is to obtain a high harvest of PEs (Delgado–Frias *et al* 1988). In order to accomplish this, a combination of software and hardware is required. The knowledge base (represented as a semantic network) is mapped onto an irregular wafer–scale array. Mapping a semantic network onto the WSI array requires several factors to be taken into account; these are: node connectivity, PE internal status, node position in the semantic network, input/output requirements and PE communication capabilities. Nodes with few connections, or near to the edges of the semantic network, may be allocated to PEs with faulty neighbours. On the other hand nodes with a high degree of connectivity must be mapped onto areas with a good number of working PEs. Placement techniques, such as those developed for gate arrays, may be used to map the semantic network onto the silicon array.

The semantic network of a scene labelling system, shown in Figure 7, is mapped onto a two–dimensional 3 × 3 array shown in Figure 8. In this small array the three faulty PEs are avoided, buffers are used to provide the connectivity required, and only nearest neighbour communication is allowed.

The robustness of semantic network algorithms allows one to deal with physical defects; that is inferences in this machine can be carried out despite failures at running time. However, the measure of certainty will decrease since information from a node or link cannot be retrieved; this is discussed below.

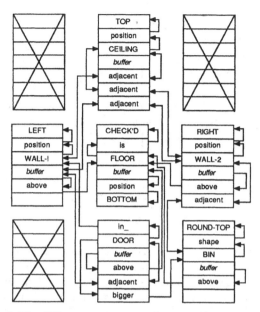

Figure 8 The KB semantic network mapped onto the 2–D array

MATCHING IN THE SN ARCHITECTURE

The SN architecture operates in a data parallel fashion; nodes that are entered in the machine must be compared with the knowledge base. This process is performed at the same time in all the machine nodes. In order to determine the value of a variable (or input node) it is necessary to match this variable and its links with nodes in the knowledge base (KB). This process is accomplished by matching node–link pairs. A neighbour node and the link that connects it to the node that is being computed are matched. Once all the node–link pairs of the node are matched a measure of believe (MB) for the variable is obtained. If such a MB is above a threshold value the variable can be assigned, otherwise the node remain unassigned.

Figure 9 shows how a node–link pair is matched. There are three possible forms on which the *node* (from the node–link pair) can be matched: 1) the node in the machine has no variable assigned (such as node–1 in the figure); 2) the node is a fixed node (i.e. in order to match it the input node must be identical); 3) the node has a variable assigned to it (in such case the input node must be identical to the assigned variable). When a node has no assigned variable, e.g. node–1, any variable can be accepted, therefore at time *t1*, this node sends a message to the link: *any* match may be performed. In the next step at *t2* the link of the KB is compared to the input link; if there is a match the PE in charge of such link will send a message to node–0 of a *weak match* because there is no strong evidence that the neighbour's variable can be assigned to node–1. In the next step, the node–0 may receive other possible matches, this node accepts the one with strongest match. When the node in the KB is fixed or a variable have been already assigned, the input *node* must be identical to

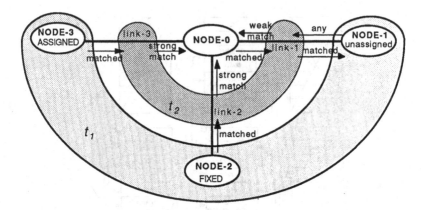

Figure 9 Node–link pair matching

the fixed value or be the same variable (this is checked at *t1*). Then the *link* of the node–link pair is matched. If a match occurs the PE sends a message to node–0: a *strong match* has been found. This process is executed in all the nodes and links of the knowledge base; thus, a parallel search for the best match is accomplished.

SCENE LABELLING

Among the applications of semantic networks is *high level* computer vision where semantic networks are used to infer what is in a scene. Scene labelling using semantic networks is performed with a finite set of labels, objects, relations and constraints (Ballard and Brown 1982). A knowledge base (KB) is stored in the array; such a KB contains information about objects, relationships among them, and their features. Figure 7 shows a semantic network KB of a room. Each link has a weight which depends on the uniqueness of the link, the extraction capabilities (how easy it is to obtain this feature from a vision system), the weight effect on the neighbours, the repeatability and any common features with other objects. Figure 8 shows the mapping of the semantic network onto a 3×3 array.

The machine input is a semantic network that describes the actual scene. Each node of the scene is matched with the KB in parallel; a measure of believe (MB) is computed to evaluate the degree of matching. The the measure of believe of a node i (MB_i) is given by

$$MB_i = \frac{\sum w_i}{N_i}$$

where w_i is the weight of the matched links that are connected to the node i and N_i is the node's highest expected weight. Once a node has a MB higher than a threshold value the variable is assigned to that node as a potential match. Assignments to other nodes may support or reject such assignment; in the latter case the MB is affected. Figure 10 shows a scene that is to be labelled; the semantic network that describes

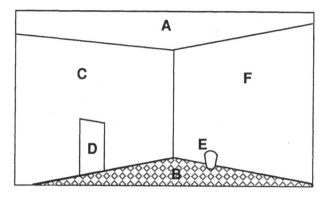

Figure 10 Scene to be labelled

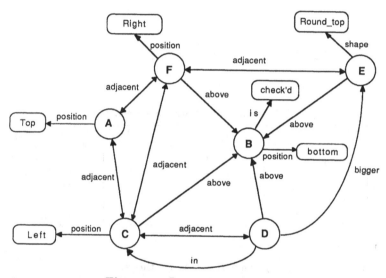

Figure 11 Scene semantic network

it is shown in in Figure 11.

The following results are obtained, when the scene semantic network (Figure 11) is matched to the KB semantic network (Figure 8). Each input node is matched to each node of the KB semantic network; MB for each node is computed. The threshold value was set to 65. If a perfect match exists the MBs shown in Table 1 are obtained.

Final assignments: ceiling (A 100); wall–one (C 100); wall–two (F 100); door (D 100); bin (E 100); floor (B 100).

The robustness of these algorithms allows to have missing objects or links. The semantic network missing parts may be due to noise in the picture, blocked objects, or to faults in the wafer–scale system. When an object is missing, such as the bin, the MBs is shown in Table 2. Final assignments: ceiling (A 100); wall–one (C 100); wall–two (F 89); door (D 68); bin ( ); floor (B 94).

Table 1 Match with a perfect semantic network

Node	CEILING	WALL-1	WALL-2	DOOR	BIN	FLOOR
A	100	21	21	0	13	0
B	0	0	0	0	0	100
C	44	100	26	15	20	0
D	0	5	5	100	6	0
E	22	15	15	15	100	0
F	44	36	100	15	6	0

Table 2 Match with a missing object

Node	CEILING	WALL-1	WALL-2	DOOR	BIN	FLOOR
A	100	21	21	0	13	0
B	0	0	0	0	0	94
C	44	100	26	15	20	0
D	0	5	5	68	6	0
F	44	36	89	15	6	0

Table 3 Match with missing links

Node	CEILING	WALL-1	WALL-2	DOOR	BIN	FLOOR
A	77	10	10	0	13	0
B	0	0	0	0	0	100
C	44	100	26	15	20	0
D	0	5	5	100	6	0
E	0	5	5	15	86	0
F	22	36	89	15	6	0

When several links are missing the MB decreases; the results obtained are shown in Table 3. Final assignments: ceiling (A 77); wall-one (C 100); wall-two (F 89); door (D 100); bin (E 86); floor (B 100).

From the last two results it is observed that the MB is affected by the missing components of the semantic network but the system can still make inferences about the objects.

The computer vision algorithms on serial machines would take a time proportional to the number of nodes that have to be visited. Some search-guiding heuristics may help to reduce the amount of searching but they must be carefully handcrafted to handle a particular set of problems. The computational complexity in a sequential machine may be a high as $O(N \times B^L)$ where N is the number of nodes in the knowledge base, B is the branching factor and L is the length of the pattern. Algorithms that use local knowledge (i.e. based on local connections) are best suited to the semantic network. The proposed semantic network architecture would require a time proportional to the number of input nodes $O(B \times L)$, since branches are searched in parallel.

CONCLUSION

A WSI 2–D array semantic network architecture may have an impact in the field of computer vision. Such an architecture could provide not only the speed, reliability and parallelism needed but also the compactness that is essential for some applications such as mobile robots. This architecture has a defect–avoidance and fault–tolerant scheme in order to overcome the WSI problems. Link weights provide support for scene labelling tasks by identifying unique features so that the matching process is faster. The target size of the array is 32×32, but much larger semantic networks can be manipulated since each PE may handle several nodes and links.

ACKNOWLEDGMENT

This work was undertaken as part of the UK Alvey Programme (Project ARCH019) whose support is gratefully acknowledged.

References

Ballard, D. H. and Brown, C. M., *Computer Vision*. Englewood Cliffs, NJ: Prentice–Hall, Inc, 1982.

Brachman, R. J., "On the Epistemological Status of Semantic Networks," in *Associative Networks*, N. V. Findler (ed), New York, NY: Academic Press, pp. 3–50, 1979.

Delgado–Frias, J. G. and Moore, W. R., "Parallel Computer Architectures for AI Semantic Network Processing," *Knowledge–Based Systems*, vol. , pp. , 1988.

Delgado–Frias, J. G., Moore, W. R. and Trotter, J. A., "High Harvest Approaches for 2–D Arrays," in *Yield Modelling and Defect–tolerance in VLSI*, W. R. Moore, W. Maly and A. J. Strojwas (ed), Bristol, England: Adam Hilger, pp. 191–202, 1988.

McDonald, J. F., Rogers, E. H., Rose, K., and Steckl, A. J., "The Trials of Wafer–Scale Integration," *IEEE Spectrum*, vol. 21, pp. 32-39, October 1984.

Peltzer, D. L., "Wafer–Scale Integration: The Limits of VLSI?," *VLSI Design*, vol. 4, pp. 43-47, September 1983.

Stapper, C. H., Armstrong, F. M., and Saji, K., "Integrated Circuit Yield Statistics," *Proceedings of the IEEE*, vol. 71, pp. 453–470, April 1983.

Uhr, L., *Multi-computer Architectures for Artificial Intelligence*. New York, NY: John Wiley & Sons, 1987.

Wah, B. and Li, G–J., *Computers for Artificial Intelligence Applications*. Washington, D.C.: IEEE Computer Society Press, 1986.

Winston, P. H., *Artificial Intelligence*. Reading, MA: Addison–Wesley Company, 1984.

Chapter 6

NEURAL ARCHITECTURES

Research on neural network models have been carried out for many years. Recently, many computer scientists have become interested on this topic. The models are believed to have a potential for new architectures for computing systems; such systems may be able to achieve human–like performance in some fields. In the literature, the artificial neural networks are also referred to as connectionist models, parallel distributed processing models and neuromorphic systems. Regardless of the name, these networks are based on what is known of the biological nervous system. The potential benefits of these systems include *fault–tolerance*, such that the malfunction of a few processing units does not significantly affect the overall performance. *Adaptation* or *learning* is another benefit: a neural network modifies its internal connections and weights in order to satisfy the desired output with the present input sometime during a specific training stage. By building artificial neural networks we may also generate a better understanding of the biological networks.

ARCHITECTURES BASED ON HOPFIELD NETWORK MODEL

In this chapter four neurocomputer architectures are presented. Most of these machines are based, in some degree, on the Hopfield network model (Hopfield 1982). This model presents several attractive features such as a good theoretical background, a regular and simple structure, and simple and local learning algorithms.

Faure and Mazaré §6.1 have proposed a highly parallel architecture based on an asynchronous cellular array. Communication between cells is accomplished by means of a hardware message routing mechanism that uses relative displacement. A cell is divided into three functional blocks: recognition, learning and routing. Direct integration of this architecture might not be possible with today's VLSI technology; the technology constraints and the schemes to overcome them are discussed in the paper.

The paper by Weinfeld §6.2 presents a system that implements a Hopfield network. The complexity of both architecture and learning algorithm are overcome by implementing *temporal connectivity*. The inclusion of learning on chip allows some local auto–adaptivity, a higher level neuromimetic property. The network state is stored in a N×1 bit shift register in order to simplify signal routing and complexity.

Ae and Aibara §6.3 describe an orthogonal neurocomputer architecture that is used to solve combinatorial problems that are represented in a matrix form. The architecture consists of Hopfield neural networks and a conventional uniprocessor computer. A possible implementation of the system in 3–D VLSI technology is shown.

Pattern and image recognition carried out by an associative memory must cope with spatial shifts. A shift–invariant neural network is studied by Prados and Kak §6.4. This network is a slight modification of the conventional neural network to provide shift invariant pattern recognition. Neurons are limited to neighbourhood interconnection which significantly reduces the number of connections. This network can be used as a low–level classifier; small objects can be recognized regardless of their location in the image.

References

Hopfield, J. J., "Neural Networks and Physical Systems with Emergent Collective Computational Abilities," in *Proc. of the National Academy of Sciences*, pp. 2554–2558, USA, 1982.

6.1 A VLSI IMPLEMENTATION OF MULTILAYERED NEURAL NETWORKS

Bernard Faure and Guy Mazaré

INTRODUCTION

Recent research in image processing, speech synthesis and word recognition by neural networks has pointed out some encouraging results. Software simulations of neural networks are so time consuming when run on conventional computers, there is no way out to increase performances but hardware implementations. Fortunately, compared to a computer, a modelised neuron is quite limited in its processing capabilities with regard to its speed, the information it acts upon and the information it produces.

To perform efficient simulations of usable neural networks, i.e. composed of a large number of interconnected neurons, we can take advantage of VLSI parallel architectures. Indeed, the inherently parallel nature of neural networks permits high computation rates, even for solving problems which cannot be processed on regular computers.

Neural networks typically provide a greater degree of robustness or fault tolerance than any Von Neumann sequential computer because of their many more processing nodes, each with primarily local connections. Damage to a few nodes or links thus need not impair overall performance significantly.

Learning from examples rather than from mathematical methods is one of the most powerful and useful features of neuromorphic systems. Users do not have to write complex algorithms neither to know how the neural network will provide a solution. According to this, a neural network creates its own internal representations needed for a given problem.

THE NEURAL NETWORK MODEL

The main interest of neural networks is their abillity to adapt their topology to the environment when they are trained (Hinton 1987). Among all the existing neural net models (Lipmann 1987), the Minsky and Papert's Perceptron is not powerful enough, the model of Carpenter and Grossberg has to deal with too complex algorithms, some others have associated learning rules minimizing global variables not easy to handle, Hopfield's have regular but very high connectivity requirements and for those reasons are not well suited for being implemented efficiently on a VLSI device. To pass these limits, Le Cun (1987) and Rumelhart *et al* (1986) have proposed the multilayered neural network model and the back-propagation learning algorithm.

The model of the formal neuron introduced more than 40 years ago by McCulloch and Pitts is still used in all the today's neural networks. This modelized neuron is made of four major components (Figure 1) :
- input connections (synapses) through which the neuron receives activation from other neurons in the network,
- a summation function that combines the various input activations into a single one,
- a threshold function that converts this sum of input activations into an output activation,
- output connections (axonal paths) by which the output activation of a neuron arrives as input activation at other neurons in the network.

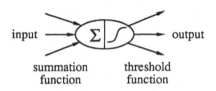

$$\text{input} \longrightarrow \boxed{\Sigma \mid \int} \longrightarrow \text{output}$$

summation   threshold
function   function

Figure 1 Formal neuron

Recognition dynamic

A neuron j of the network is well-defined by its sigmoidal non linearity f which converts its input activation a_j (equation 1) into an output activation x_j (equation 2) :

$$a_j = \sum_i w_{ij} x_i \tag{1}$$

$$x_j = \alpha \; \frac{e^{\beta(\sum_i w_{ij} x_i)} - 1}{e^{\beta(\sum_i w_{ij} x_i)} + 1} \tag{2}$$

If θ_j is the threshold of the neuron j, we can assign it to a weight held by a particular connection which activation is always 1. This is done by creating an extra neuron 0 with an input from the outside world and a connection to each unit of the neural network, weighted by $(-\theta_j)$. The threshold inverse is now replaced by a term $w_{0j} x_0$ and is part of the neuron input activation. The synaptic weight w_{ij} modulates the activation passing through the connection from neuron i to neuron j and α and β are constant parameters.

In the multilayered model, each neuron of a layer is fully connected to all the neurons of its two adjacent layers (Figure 2). In recognition mode, the network propagates concurrently the output activation of all the neurons of a given layer to the neurons of the next downstream connected layer.

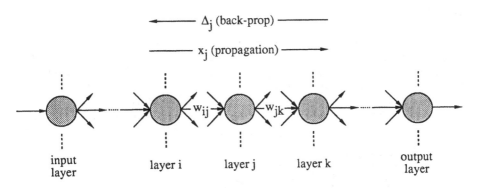

Figure 2 Multilayered network

Learning rule

In learning mode, the network back-propagates the local gradients computed by each neuron of a layer to all the neurons in the previous upstream connected layer. Each neuron updates the weights of its input connections in a way that decreases the error on its own output activation.

For a neuron of the output layer s, the local gradient Δ_s depends directly on the difference between the network output and the output expected by the outside world :

$$\Delta_s = (S - Y) \, f'(a_s) \tag{3}$$

$$w'_{ks} = w_{ks} - \gamma \Delta_s x_k \tag{4}$$

S is the current network output and Y is the desired output, both related to the same input pattern applied to the neural net. a_s is the neuron input activation and f' is the derived of the sigmoidal threshold function f. w_{ks} is the synaptic weight associated to an input connection of a neuron in the output layer, x_k is the activation passing through the considered connection. γ is a constant parameter determining the gain of one step of the learning algorithm.

For a neuron of the hidden layer j, the local gradient Δ_j is only related to the gradients Δ_k back-propagated by the neurons of the adjacent - downstream connected - layer k.

$$\Delta_j = \left(\sum_k w_{jk} \Delta_k \right) f'(a_j) \tag{5}$$

$$w'_{ij} = w_{ij} - \gamma \Delta_j x_i \tag{6}$$

The updated synaptic weight w_{ij} is associated to an input connection of the neuron.

THE CELLULAR ARRAY

To implement such neural networks, we propose the use of a message-based communicating cellular array, which has previously been studied for several distinct applications and which has already been presented under its logical simulator form (Objois *et al* 1987). This array consists of an N x N matrix of asynchronous cells. Each cell is physically connected to its four immediate neighbors through eight unidirectional buffers (Figure 3). A flip-flop based mechanism included in each buffer prevents multi-access from adjacent cells.

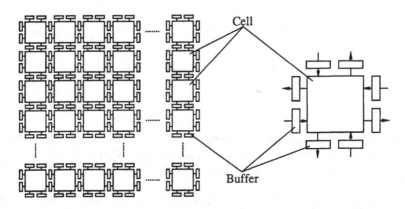

Figure 3 Asynchronous cellular array

Message transmission

The main original feature of this architecture is the ability for each cell to be logically connected to any other cell of the array and to the outside world by using a very simple but efficient hardware message routing mechanism, distributed among the cells. Each message must supply the cell with all the information needed for its processing (Figure 4).

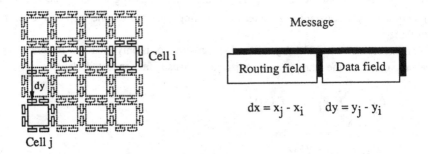

$$dx = x_j - x_i \qquad dy = y_j - y_i$$

Figure 4 Message transmission

A message must hold a data field containing the information to be processed by the application implemented on the array and, to be handled by this communication system, a routing field with information needed by the array for its own use. This routing field must include at least the *relative displacement* dx and dy from the current to the addressee cell (i.e. the number of cells the message has to cross to reach its destination).

Before we can use this architecture for computations, we have to initialize it with the map of the neural net we want to simulate. As the only way to send information to a cell is by message transfer, we must send a set of initialization messages to each cell of the array. The message itself must tell its destination cell what it has to do with the transmitted data by using particular flags in its routing field. The message structure is as follows :

routing field :
- *relative displacement* dx and dy to the addressee cell
- *internal memory address* in the addressee cell : specifies a particular link
- *type* : describes the data field contents
    - dx
    - dy
    - internal memory address
    - cell function (i.e. summation and threshold, summation only, or no operation)
    - initial synaptic weight
    - neuron activation
    - back-propagated gradient

data field :
- *value* : for the use of the destination cell (according to the type)

Cell function

The cells must perform the computations needed by the model of the formal neuron and must also handle the inter-cell connections, which reflect the neural net topology, by storing the synaptic weights and the initial routing values for each link. Thus, a cell is functionally divided in two concurrent parts : a processing and a routing part (Figure 5).

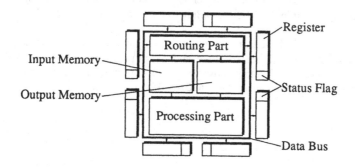

Figure 5 Cell functional diagram

The unique data bus surrounding the cell is managed and accessed only by the routing part. The processing part can only access its input buffer and output memory in which the routing part writes inputs and reads outputs. If we permit the recognition and learning processes to occur simultaneously in a cell, we can divide its processing part in a recognition part and a learning part which communicate between themselves through a shared memory (for weight values) handled in the same way as the external buffers.

We have chosen the cell starting a connection to handle its associated synaptic weight in order to keep the internal arithmetic resolution of a cell to a high enough order without having to transmit wide messages during the learning phase (the message holds Δ_k and the learning part computes internally $w_{jk}\Delta_k$) and to be able to manage inter-neuron connections easily. This is equivalent to the model because the weights are updated with the incoming gradients and not with the computed local ones which are transmitted directly to the neurons of the upstream connected layer.

Routing process. After inspecting all its input buffers to acquire an incoming message, the routing part determines the output buffer to which it has to transmit the message. This output buffer can be an input buffer of a neighboring cell or the input buffer of the cell processing part. If the transmission is possible (selected output buffer empty), the message is processed, otherwise, it is held in the input buffer. The OCCAM-like algorithm is :

```
SEQ
    select the input buffer of highest priority
    WHILE TRUE
        SEQ
            IF the selected input buffer is full
                SEQ
                    read the message
                    according to its routing field, select an output buffer
                    IF the selected output buffer is empty
                        SEQ
                            update the routing field and write the message to the output buffer
                            mark the output buffer as full and the input buffer as empty
            select the next input buffer according to the predefined priority
```

Recognition process. Each neuron computes its output activation by summing all its incoming activations until all are received and passing the result through the sigmoid function. It stores its weighted output activations in its output memory for being handled by the message transmission mechanism. The OCCAM-like algorithm is :

```
WHILE TRUE
    IF the input buffer is full
        SEQ
            read the input buffer and mark it as empty
            IF the incoming message holds an initialization value
                according to the type and internal address
                initialize dx, dy or address memories or cell function or weight memory
            IF the incoming message holds a weighted neuron activation
                SEQ
                    mark its flag as received and sum it to the internal sum variable
                    IF all messages arrived
                        SEQ
                            apply the sigmoid non linearity on the sum variable
                            FOR EACH output logical connection
                                SEQ
                                    weight the obteined output activation
                                    store the result in its position in the output memory
                            clear the sum variable and mark the flags with no message received
```

Learning process. Each neuron waits for all the local gradients sent by the downstream-connected neurons, computes its own gradient which it sends to the upstream-connected neurons by storing it in its output memory and updates the weights it handles with the incoming gradients. The OCCAM-like algorithm is :

```
WHILE TRUE
    IF the input buffer is full
        SEQ
            read the input buffer and mark it as empty
            IF the incoming message holds an initialization value
                according to the type and internal address
                initialize dx, dy or address memories or cell function or weight memory
            IF the incoming message holds a local back-propagated gradient
                SEQ
                    store it in its position in the internal memory
                    WHEN the internal memory is full
                        SEQ
                            perform the weighted sum of all the values in it
                            generate the local gradient and store it in the output memory
                            FOR EACH synaptic weight
                                update it with its related gradient stored in the internal memory
                            clear the internal memory
```

ENVIRONMENT

The cellular array is a specific device which must be connected to a host computer able to manage the array inputs/outputs through a link interface. The host computer has to initialize the array with a map of the neural net before it can do any computing, and once this has been completed, it has to control the simulation. The OCCAM-like algorithm is :

```
SEQ
    create the neural network map
    send the initialization messages to the array
    PAR
        send the input messages to the array (patterns to be recognized or learned)
        get back the output messages
    UNTIL end of simulation
    edit results and network statistics
```

VALIDATION

We have described and functionally simulated a 6 x 6 cellular array with link interfaces and host computer. The simulator of the whole system has been written in OCCAM and run on a B004 Transputer board (consisting of one IMS T414 and 2 megabytes memory) connected to a PC-AT. In this simulator, each physical entity (i.e. host computer, link interface, buffer, routing part, recognition part and learning part) is described by an OCCAM process and all the data transfers and synchronization are handled by communications through OCCAM channels.

Simulations have been performed for an exclusive-OR and for "Little Red Riding Hood" (Jones and Hoskins 1987). These applications required respectively 5 and 16 neurons, with one more for handling the thresholds. We are currently studying the behavior of a neural net dedicated to an application of pattern recognition which needs about 80 neurons.

First simulations were performed using all the computational power of the IMS T414 Transputer and then, several modifications were made to simplify drastically the operations and so take into account the cell complexity for a VLSI implementation.

INTEGRATION

The way the array manages the connections demands a quite limited connectivity for a cell. Eight input and eight output connections seem to be a good figure for the surface used. To deal with this constraint, one neuron can be assigned to a group of cells :
- an input sub-tree in which each cell sums its weighted inputs received from other neurons in one partial sum and outputs it
- a cell which sums the partial sums arriving from the input sub-tree, passes the result through a non linearity and sends it to the output sub-tree
- an output sub-tree in which each cell receives one input from the central neuron cell, weights it and outputs the results to other neurons

Our goal is to integrate the maximum number of elementary cells on a single chip ; with the perspective of simulating large neural nets able to perform simultaneously several tasks, each devoted to a particular sub-net.

The main problem of the VLSI implementation is to limit the cell complexity and the best way is to limit its computing complexity by avoiding all floating-point calculations and sophisticated functions. Thus, we have tried to replace the continuous sigmoid function and its derived with a linear interpollation (between the four values 0, 1, 2 and 10) and we have found that the network needs 20% more learning time to give comparable results but that the learning process still converges (Figure 6). These curves show a neuron activation on y-axis, versus the number of iterations on the x-axis. The convergence is obtained when the output is near +1 or -1. By reducing the cell internal data resolution down to a 10-bits wide, the learning process still converges with another 20% more time needed. With a lower resolution, we found no convergence for the trained examples.

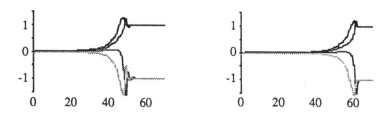

Figure 6 Network output activations

Of course, we could not integrate an array with thousands of cells on a single chip today, though future advances in technology may make this possible. We can easily overcome this limitation by developing a PC-board with an array of chips, each one realizing a small array or by exploiting Wafer Scale Integration.

This cellular architecture can be improved with a class of routing algorithms which solve the problem of wafer reconfiguration (Faure *et al* 1986). When a wafer is tested and the defective cells located, the array is loaded with the defect map by indicating to each valid cell the status (valid or not) of each of its four neighbors. The routing algorithm is then able to re-route messages through the array, and let them dynamically find a path avoiding defective cells, up to their destination.

CONCLUSION

There are two possibilities for increased performance of standard architectures : to keep the Von Neumann concept with a more advanced processor able to perform complex operations or to take advantage of multiprocessor architectures such as cellular arrays where the work is distributed over a large number of simple processors.

A neural network is well-defined by the distributed algorithms performed by a neuron and the interactions between neurons in the net. The underlying parallelism is evident with the association neuron-processor which first comes in mind.

Many applications are good candidate for being processed efficiently by such a cellular array. We have designed and evaluated some placement and routing algorithms and we are currently working on parallel image reconstruction (Lattard and Mazaré 1988), logical simulation and data flow processing.

Our goal is to point out the superiority of a parallel architecture using this communication technique over a systolic architecture and over a standard Von Neumann computer for a large class of applications.

ACKNOWLEDGEMENT

This work is supported in part by the *Pôle Architecture* of the group *Cooperation, Concurrence et Communication* (C^3) of the *Centre National de la Recherche Scientifique*.

REFERENCES

Faure, B., Ansade, Y., Cornu-Emieux, R. and Mazaré, G., "WSI Asynchronous Cell Network", in *Wafer Scale Integration*, G. Saucier and L. Trilhe (Eds.), Grenoble France : North Holland, 1986.

Hinton, G. E., *Connectionist Learning Procedures*, Technical Report CMU-CS-87-115, Computer Science Department, Carnegie-Mellon University, Pittsburg Pa, June 1987.

Jones, W. P., Hoskins, J., "Back-Propagation : a Generalized Delta Learning Rule", in *BYTE Magazine*, pp. 155-162, October 1987.

Lattard, D. and Mazaré, G., "Parallel Image Reconstruction by using a Dedicated Asynchronous Cellular Array", in *Proceedings of the International Conference on Parallel Processing for Computer Vision and Display*, Leeds England, January 1988.

Le Cun, Y., *Modèles Connexionnistes de l'Apprentissage*, Thèse de Doctorat, Spécialité Informatique, Université de Paris 6, juin 1987.

Lippmann, R. P., "An Introduction To Computing with Neural Nets", in *IEEE Acoustics Speech Signal Processing Magazine*, n° 4, pp. 4-22, April 1987.

Objois, P., Mazaré, G., Cornu-Emieux, R. and Ansade, Y., "Highly parallel logic simulation accelerators based upon distributed discrete-event simulation", in *Proceedings of the International Workshop on Hardware Accelerators*, Oxford England, October 1987.

Rumelhart, D. E., Hinton, G. E. and Williams, R. J., "Learning Internal Representations by Error Propagation", in *Parallel Distributed Processing : Explorations in the Microstructure of Cognition*, vol. 1 : Foundations, D. E. Rumelhart and J. L. McClelland (Eds.), Cambridge Ma : MIT Press, 1986.

6.2 A FULLY DIGITAL INTEGRATED CMOS HOPFIELD NETWORK INCLUDING THE LEARNING ALGORITHM

Michel Weinfeld

INTRODUCTION

Man has since long tried to understand the way the mind works, and how it can exhibit such extraordinary capacities, especially when compared to the machines and processes which have been built in order to try to imitate even the simplest of its features, such as pattern recognition (optical and acoustical), sensori-motor coordination, etc.

As progress is made in a better understanding of global and detailed brain mechanisms, new ideas and concepts emerge which could help give more legitimacy and more meaning to the term "Artificial Intelligence".

It has been seen by simulation, and sometimes demonstrated that the so called "connectionist architectures", inspired by the real cortex structures and the formalization of the neuron, can lead to properties mimicking some of the mind's elementary properties, for instance generalization, pattern classification, associative memory. These architectures are not programmed, but they learn. They are made of numerous elementary processors (the neurons) which are individually capable of a low level task, namely they integrate several stimuli and take an almost binary decision as a result of this integration. What makes them have an interesting collective behaviour is their connections, that is the way they exchange elementary information. And, last but not least, they are robust with respect to connections and cell degradations or even failure, hence reminding us also of a capital biological property.

Among the various architectures appearing in the literature, the fully connected network suggested by Hopfield (1982) has many very interesting features:

- the increasingly good theoretical knowledge about its properties, which are those of a dynamic system, tractable by analogy with spin glass statistical mechanics,
- a maximally regular and simple structure,
- a very simple operating mechanism at the neuron level: flipping based on a weighted majority, implementing "easy" arithmetic and/or logic,
- the existence of simple and local learning algorithms,
- its intrinsic ability to perform content addressable memory function, making it interesting for such engineering purposes as signal processing, error correction, etc...
- possibly representing a basic building block for higher level neuromimetic structures.

We shall review some of these properties more in detail later. Nevertheless, one can understand quickly that the fully connected networks seem to be very well adapted candidates for VLSI integration. This integration is primarily justified as a help to research: numerical simulations, even performed on increasingly powerful and sophisticated computers, tend to become a bottleneck. If one wants to investigate higher hierarchical structures (networks of networks, having probably more operative properties than today's networks), the availability of integrated circuits exhibiting some fundamental characteristics such as associative memory or classification-generalization may help. On the other hand, there is little doubt that the forthcoming applications of artificial intelligence in general, and especially connectionist structures, will depend rather strongly on the existence of dedicated chips. These theoretical as well as practical reasons are the motivations for the work which is presented here, as they are without doubt for other similar projects, already described or to be published (for instance Sivilotti *et al* 1984, Schwartz *et al* 1986, Alspector *et al* 1987, Moopenn *et al* 1987).

THE PRINCIPAL PROPERTIES OF FULLY CONNECTED NETWORKS

Dynamics

The decision making rule of a formal binary neuron is as follows: if the weighted sum of all the inputs from the others neurons is positive, the neuron becomes or stays active; its state is represented by the variable σ, the value of which is +1. In the other case, it becomes or stays inactive, $\sigma = -1$. This individual mechanism is performed asynchronously by all neurons together. At every moment t, the state of the whole network can be represented by a vector $\underline{\sigma}$, the components of which being the state of each neuron. The connection between any two neurons i and j is represented by the so-called synaptic coefficient C_{ij}, its value being calculated by a learning algorithm. In fact, the asynchrony can be relaxed to a partial or even total synchronous updating of the neurons, provided that the learning rule is not too exotic, as shown by numerical simulations. In this case, the dynamics of the network is simply given by the expression:

$$\underline{\sigma}\,(t+\tau) = \text{sign} [\, C \, \underline{\sigma}\,(t) \,] \tag{1}$$

which can also be put in scalar form:

$$v_i = \Sigma_j C_{ij} \, \sigma_j(t) \;, \;\; \sigma_i(t+\tau) = \text{sign} \, (v_i) \quad, \quad \text{i and j varying from 1 to N} \tag{2}$$

giving the updated state vector at time $t+\tau$ as a function of the previous state vector and of the synaptic matrix C. This simple rule allows to derive an energy function, the minima of which indicates temporally stable states of the network, called attractors. These attractors depend on the way the synaptic matrix has been calculated. The associative property of the network lies in the fact that if one initializes it, by forcing its state into a certain vector, it will generally dynamically converge to the attractor state which is the closest (in terms of Hamming distance) to this vector. The speed of convergence is independent of the size of the network (Peretto and Niez 1986). Of course, the main issue is in finding the appropriate rule to build a connection matrix providing the best properties, for instance maximizing the number of stored attractors, allowing them to be partially correlated, etc.

The learning algorithm

Vectors to be learned (prototypes) are presented to the network in sequence, the synaptic coefficients are computed iteratively. The simplest rule is called Hebb rule, it can be stated in a simple expression:

$$\Delta C_{ij}(k) = \sigma_i^k \sigma_j^k / N$$

where k indicates the number of the current prototype being learned and N the number of neurons. This rule is good with respect to dynamics and general properties of the networks. Unfortunately, it generates many parasitic attractors when the prototypes are not strictly uncorrelated from each other, and gives a limited capacity in term of the ratio α of the number of prototypes that can be learned and retrieved, to the number N of neurons (about 14%) (Hopfield 1982, Amit *et al* 1985).

Another rule has been suggested by Personnaz *et al* (1985, 1986), it is called the projection rule. This rule considers p prototypes { $\underline{\sigma}^1$, $\underline{\sigma}^2$, ... $\underline{\sigma}^p$ }, and calls Σ the matrix made of the column prototype vectors. The projection learning rule consists in computing the synaptic matrix C by the relation:

$$C = \Sigma \, \Sigma^I$$

where Σ^I is the pseudoinverse of matrix Σ. The C matrix is the orthogonal projection matrix in the subspace spanned by the prototype vectors. It provides a learning ratio $\alpha = p/N$ close to unity, and tolerates very well partially correlated prototypes. This represents improvements to Hebb rule. It can be expressed in matrix form:

$$\Delta C(k) = C(k) - C(k-1) = (\underline{\sigma}^k - \underline{v}^k)(\underline{\sigma}^k - \underline{v}^k)^t / \| \underline{\sigma}^k - \underline{v}^k \|^2$$

where $\underline{\sigma}^k$ is the kth prototype presented, and $\underline{v}^k = C(k-1) \, \underline{\sigma}^k$

Figure 1 shows an example of how this rule works. It shows the histogram of 1000 retrievals with stimulus at initial normalized Hamming distance H_i from the considered prototype state, and for a learning ratio α. H_f is the final normalized Hamming distance after convergence. The network has a size N of 64 neurons.

It can be seen that the computation of any coefficient needs to evaluate the *global* quantity represented by the denominator. This is not a problem for numerical simulations, but is a drawback for the implementation of a circuit, since it makes it more complex than in the case of a *local* rule. For that reason, several approximations of the previous expression can be derived:

$$\Delta C_{ij}(k) = (\sigma_i^k - v_i^k)(\sigma_j^k - v_j^k) / N \tag{3}$$

$$\Delta C_{ij}(k) = (\sigma_i^k \sigma_j^k - v_i^k \sigma_j^k - \sigma_i^k v_j^k) / N \tag{4}$$

$$\Delta C_{ij}(k) = (\sigma_j^k - v_j^k) \sigma_i^k / N \tag{5}$$

$$\Delta C_{ij}(k) = (\sigma_i^k - v_i^k) \sigma_j^k / N \tag{6}$$

These formulas have various complexities and properties, but share the same property of locality. As regards formula 6, it has been shown that its repeated application with the same set of prototypes leads generally to a matrix which is the projection matrix (Diederich and Opper 1987). Thus, with an increased complexity in time, the global rule can be recovered with only local implementation.

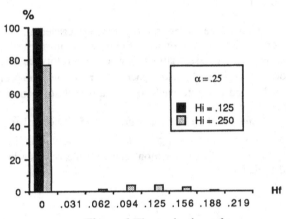

Figure 1 The projection rule

There is no mean to predict the exact number of repeated presentation of the set of proto-types needed to reach the projection matrix. This means that one has to wait until the C_{ij} have stabilized, provided that they are defined with a sufficient precision. The simulations show, however, that this stabilization is rather fast. Figures 2 and 3 show that this precision has a visible influence on the quality of the retrieval. The conditions and variables are the same as on Figure 1, but the precision of the coefficients is voluntarily limited to **m** bits. All these results, due to Dreyfus *et al* (1988), have to be taken into account when designing an inte-grated circuit, and have a strong influence on the total number of transistors, because there are N^2 m-bits synaptic coefficients to implement.

THE CIRCUIT ARCHITECTURE

An important choice: analog or digital ?

Most of the published designs describe analog implementation of fully connected networks, following ideas of Hopfield (1984). These designs use operational amplifiers as neural cells, and resistances as connecting links. In this case, the corresponding conductances are the synaptic coefficients. It is difficult to make them variable, that is to make the network capable of learning, but this simplicity allows easy solutions to the connectivity problem. These circuits also exhibit some of the current disadvantages of analog circuits, for instance electrical instability and noise sensitivity, added to a certain difficulty to precisely define internal states and to interface with the often digital outside world.

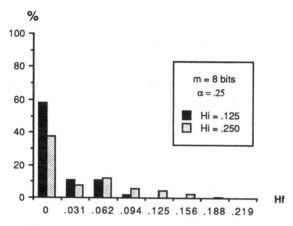

Figure 2 Limited precision projection rule, **m**= 8 bits

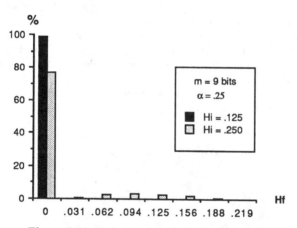

Figure 3 Limited precision projection rule, **m**= 9 bits

Generally, the designs implement sign programming (-1, 0 ,+1) by use of switches. Even if the basic properties of the networks are conserved, this kind of "synaptic plasticity" is not sufficient for practical purposes. There are current projects which try to implement the coefficients by storing charges in capacitors, the corresponding voltage driving transistors whose transconductance represents the synaptic coefficients (Schwartz and Howard 1988, Tam *et al* 1988). This solution is interesting but rather technology dependent, and does not allow for a great precision on the coefficients, owing to charge leaks and geometric tolerances.

On the other hand, fully digital implementations seem to be the answer to flexibility, ease of programmability, and precision. For instance, the circuit designed by Murray *et al* (1987) is a very elegant solution, using bit-serial arithmetic. However this circuit still needs to have its coefficients computed outside the network, making it dependent of a host processor.

Since we hope to build architectures composed of basic networks, we want to implement learning inside the network itself. Choosing the approximated rule (equation 6), one can see that the arithmetic involved is not much more complicated than the arithmetic required to implement the retrieval (dynamic) phase (equation 2), especially if one remembers that the σ have the values -1 or +1: all the multiplications are simply identities or complementations, the whole arithmetic being integer. Also, if we choose a network size which is a power of 2, the divisions are only shifts. The remaining problem is to implement the required connectivity.

The basic network architecture

With current technologies, it seems impossible to realize a chip with all the needed connections and updating circuitry working in parallel. A compromise must be made. When performing numerical synchronous simulations, one uses two nested loops of size N. It is possible to parallelize one of these loops in the silicon, keeping the other sequential. The network will include N identical neuron cells, each one with full arithmetic capability for learning and updating, and also including a local memory containing the relevant column of the synaptic matrix, that is N coefficients. The only information that neurons have to exchange with the others is their states, coded on one bit only, thus simplifying greatly signal routing and chip complexity. This basic idea is depicted on figure 4.

The network state is stored in a circular N x 1 bit shift register. A simultaneous partial potential update in every neuron is performed at each elementary shift of the register, and after a complete revolution, each neuron can take its decision: the reloading of the shift register with the new state is done in parallel.

This architecture can be made somehow better by including each cell of the state register in one neuron, reducing even more the routing requirements.

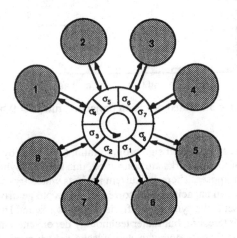

Figure 4 The basic network architecture

A more detailed view of the neuron circuitry

Figure 6 shows a symbolic schema of one neuron. The ALU includes a parallel adder, complementers, and a shifter. The "potential" v_i is stored in an accumulator, the most significant bit of which represents the new neuron state, when the updating is completed. This state is loaded in the local one-bit cell of the state register $\sigma(t)$, before a new updating revolution can take place (operation symbolized by switch S_1). It is also stored in the $\sigma(t-1)$ one-bit register, for convergence testing purpose. In effect, one supplementary updating revolution is necessary in order to test if the network is stable, that is $\sigma(t) = \sigma(t-1)$ for every neuron. This can be implemented with an exclusive-or operation simultaneously in each neuron, all the individual convergence signals ξ being or'ed in a combinatorial way. The synaptic memory is represented by $C(i)$, and contains 64 9-bits words. The way we implement this memory is by using shift registers, due to the fact that it is always addressed sequentially, in both learning and retrieval phase. This implementation may be made almost as economical (in terms of power consumption) as a classical static random access memory, despite the fact that all N^2 coefficients have to be moved at each elementary time step, instead of only the N strictly needed: the auxiliary circuitry overhead needed with a classical memory design is rather costly when the total memory capacity is low, as it is in our case. On the other hand, we refused to envision dynamic memory, which would be probably more interesting in terms of surface, to avoid design difficulties. On the schema, switch S_2 symbolizes the two modes of operation, learning and retrieving.

Notice that in learning phase, two revolutions of the state register are necessary for computing one coefficient increment: one for evaluating the potential, the other for the coefficient computation itself. The schema does not represent the circuitry for testing the coefficients convergence when using the iterative learning algorithm, but it uses the same kind of strategy that the state convergence testing: the calculated iterative increment is simply tested to zero, and all the local tests are combinatorially or'ed.

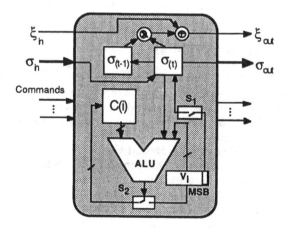

Figure 5 Simplified schematic of one neuron

The neuron sequencing is determined by the command lines which are common to the whole network, and are issued from a central clock and sequencing unit.

The geometric design of the neuron is such as the network can be designed by a tiling of identical neuron cells, with almost no routing necessary.

Figure 7 shows a conceptual sixteen-neuron network design. The I/O tasks are performed by a serial-parallel register **R**, which can be used for loading or unloading the network. It can be put out of the network loop when in learning or retrieving phase, to avoid slowing down the circuit by a factor of 2, by using the switch **S**. Wire ξ is the serial combinatorial convergence signal. For sake of simplicity, other common command signals have been omitted, as well as the sequencer, clock generator, and other ancillary circuits.

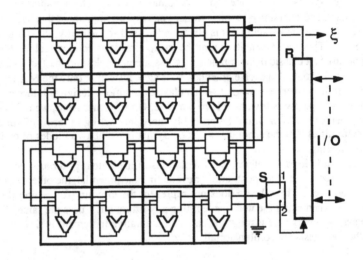

Figure 6 Schematic of a 16 neurons network

Choice of technology, design strategy

As we have made conservative choices to avoid difficulties, we shall use an existing commercial technology, namely two metal CMOS, with 2 or 1.5 μm characteristic dimension. The overall clock rate, in this case, will be around 20MHz, which will give a convergence characteristic delay of 10 to 20 microseconds. This speed is certainly interesting enough, especially when compared to the global speed of numerical simulations.

One neuron is to be designed and processed via the french multi-chip project (CMP-GCIS). It is expected to amount to 4 to 5000 transistors, which is a rather low number for testing and verification purposes. Because of the regularity of the circuit, we should be able to go on one step from a single neuron to processing the whole network, which should have a surface not far from one square centimeter. We consider as interesting the testability issue, since such a circuit can be functional even if in case of processing failures, provided that these failures are located for instance in the memory.

CONCLUSION

We have shown how it is possible to implement a fully connected network, which is known for having associative memory properties, using digital circuitry. The complexity of both architecture and learning algorithm can be dealt with by implementing "temporal connectivity" instead of trying to solve difficult (or perhaps impossible) signal routing problems on the silicon. The inclusion of learning on the chip itself will be used to build circuits (firstly on a board-level basis) which should exhibit higher level neuromimetic properties, by allowing for a certain amount of local auto-adaptivity. Larger integrated networks, if needed, will probably be made possible only by wafer scale integration (WSI), which is not yet a current technique but could benefit from the regularity and robustness properties of connectionist structures. Further studies in the VLSI domain could include improvements of the circuit speed by trying more efficient design (including dynamic circuitry), or also inclusion of more sophisticated features, for instance a temperature equivalent for improving convergence by a better elimination of spurious attractors.

Acknowledgements

This work is performed in collaboration with the Laboratoire d'Electronique, ESPCI, Paris (G.Dreyfus, A.Johannet, L.Personnaz), to whom I am greatly indebted. It is sponsored by contract 87.C.185 from Ministère de la Recherche et de l'Enseignement Supérieur, and contract 87/34/187 from Ministère de la Défense (DRET).

REFERENCES

Alspector, J., Allen, R. B., Hu, V. and Satyanarayana, S., "Stochastic learning networks and their electronic implementation", *IEEE Conf. on Neural Information Processing Systems, Natural and Synthetic*, Denver, 1987.

Amit, D. J., Gutfreund, H. and Sompolinsky, H., "Storing infinite numbers of patterns in a spin-glass model of neural networks", *Phys. Rev. Lett.*, vol. 55, no. 14, pp. 1530-1533, 1985.

Diederich, S. and Opper, M., "Learning of correlated patterns in spin-glass networks by local learning rules", *Phys. Rev. Lett.*, vol. 58, no. 9, pp. 949-952, 1987.

Dreyfus, G., Johannet, A. and Personnaz, L., private communication, 1988.

Hopfield, J. J., "Neural networks and physical systems with emergent collective computational abilities", *Proc. Natl. Acad. Sci. USA*, vol. 79, pp. 2554-2558, 1982.

Hopfield, J. J., "Neurons with graded response have collective computational properties like those of two-state neurons", *Proc. Natl. Acad. Sci. USA*, vol. 81, pp. 3088-3092, 1984.

Moopenn, A., Langenbacher, H., Thakoor, A. P. and Khanna, S. K., "A programmable binary synaptic matrix chip for electronic neural networks", *IEEE Conf. on Neural Information Processing Systems, Natural and Synthetic*, Denver, 1987.

Murray, A. F., Smith, V. W. and Butler, Z. F., "Bit-serial neural networks", *IEEE Conf. on Neural Information Processing Systems, Natural and Synthetic*, Denver, 1987.

Peretto, P. and Niez, J. J., "Stochastic dynamics of neural networks", *IEEE Trans. on Systems, Man and Cybernetics*, vol. SMC-16, pp. 73-83, 1986.

Personnaz, L., Guyon, I. and Dreyfus, G., "Information storage and retrieval in spin-glass like neural networks", *J. Physique Lett.*, vol. 46, pp. L359-L365, 1985.

Personnaz, L., Guyon, I. and Dreyfus, G., "Collective computational properties of neural networks: new learning mechanisms", *Phys. Rev. A*, vol. 34, pp. 4217-4228, 1986.

Schwartz, D. B. and Howard, R. E., "Analog VLSI for adaptive learning", *Neural Networks for Computing*, Snowbird, 1988.

Schwartz, D. B., Howard, R. E., Denker, J. S., Epworth, R. W., Graf, H. P., Hubbard, W., Jackel, L. D., Straughn, B. and Tennant, D. M., "Dynamics of microfabricated electronic neural networks", *Appl. Phys. Lett.*, vol. 50, pp. 1110-1112, 1987.

Sivilotti, M., Emerling, M. R., and Mead,C., "A novel associative memory implemented using collective computation", in *Proc. Chapel Hill Conf. on VLSI*, pp. 329-342, 1985.

Tam, S., Holler, M. and Canepa, G., "Neural networks synaptic connections using floating gate non-volatile elements", *Neural Networks for Computing*, Snowbird, 1988.

6.3 A NEURAL NETWORK FOR 3–D VLSI ACCELERATOR

Tadashi Ae and Reiji Aibara

INTRODUCTION

We propose a neurocomputer architecture with Hopfield-type memory devices for an accelerator of combinatorial problems. The architecture consists of Hopfield neural networks and a conventional von Neumann computer, and realizes the neural computation on the Hopfield neural networks interconnected with each other. We call this architecture the *Building Block Architecture*, where the 3-D VLSI technology is assumed to be applicable.

COMPUTER ARCHITECTURE WITH NEURAL NETWORK MEMORIES

In our neurocomputer model the memory part consists of a neural network memory in place of the flip-flop memory in a conventional computer (Figure 1). In general, the neural network memory includes more than one Hopfield neural networks (Hopfield and Tank 1985). We regard Hopfield neural network as the extension of the memory concept, i.e. the enlargement of a memory cell from a bi-stable circuit to a multi-stable circuit (Figure 2), and we define a Hopfield-type memory to be a fundamental block denoted by F_i ($i=1,2,\cdots,e$) in Figure 3, where the threshold logic gate in Figure 2 (c) is often replaced by a Boolean gate. In the neural

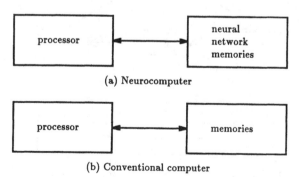

(a) Neurocomputer

(b) Conventional computer

Figure 1 Neurocomputer vs. Conventional computer

computation, the processor changes the interconnection among these fundamental blocks. As a result, the system of Figure 3 can realize the neural computation represented by a flowchart schema using several primitives as in Figure 4, where each box denoted by B means a block (defined later).

Figure 3 shows the building block architecture, that is, an architecture to realize the neural computation schema synthesized by the building block method defined as follows:

(1) A block is a fundamental Hopfield neural network for which stable vectors are known.

(2) **repeat** find perspective combination(s) of blocks,
                      and create new block(s) if necessary.
    until goal is attained.

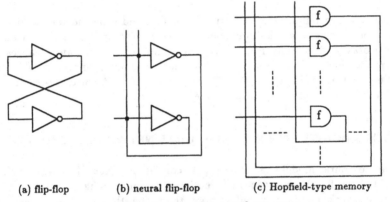

(a) flip-flop   (b) neural flip-flop   (c) Hopfield-type memory

Figure 2 Neural network memory

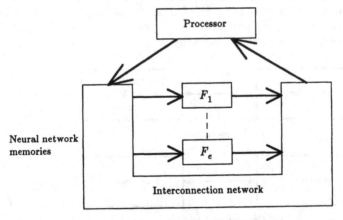

Figure 3 Extended neurocomputer configuration

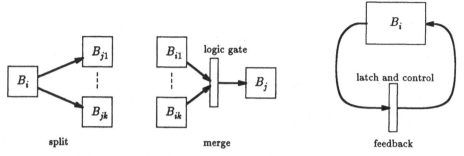

Figure 4 Primitives to construct the neural computation

Hopfield neural network is, of course, a powerful computation model. When using Hopfield's model, however, we adjust the input coefficients and the threshold of each device, to realize exactly the problem to be solved. We call it the *direct method* in the sense that the total configuration is provided in advance. This method can produce a very high-speed neural computing device since it is an asynchronous Boolean gate circuit with no pure delay, if the problem is mapped in it and if the gate circuit is directly fabricated. A large class of combinatorial problems, however, seems not to be easily mapped there, because the model is too general and difficult to be analyzed.

To improve the direct method, we introduce the building block synthesis, where each block is a Hopfield neural network, the state vectors of which have been clarified. The major difference with the direct method is that the algorithm for synthesis of blocks is introduced in the building block method. In our neurocomputer model of Figure 3, the processor plays a role of algorithms for synthesis. The direct method, the building block method and the conventional computation are compared with each other in Figure 5.

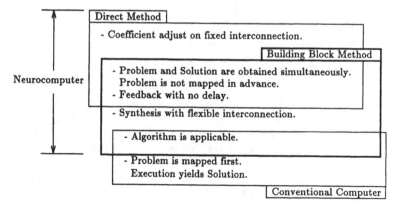

Figure 5 Comparison of Methods

FUNDAMENTAL BLOCK

For the building block method, we introduce the k-out-of-n stable circuit as a fundamental block, because its characteristics are known (Ae *et al* 1974). The k-out-of-n stable circuit (in short, $_nC_k$ stable circuit) is a Hopfield neural network since the circuit is obtained from an interconnected Boolean gate circuit, where the Boolean gate is provided by a threshold function.

The relation between the threshold and the number of stable vectors in the $_nC_k$ stable circuit is described as follows:

Threshold Function

for $y_i = f(x_1, \cdots, x_{i-1}, x_{i+1}, \cdots, x_n)$

$$
y_i = \begin{cases}
0 & \text{if } \left(\sum_{j=1}^{n} a_j x_j \right) - \theta > 0 \\[2em]
1 & \text{if } \left(\sum_{j=1}^{n} a_j x_j \right) - \theta < 0 ,
\end{cases}
$$

where $a_i = 0$ and $a_j = 1$ or 0 (for $j \neq i$), and $\theta > 0$.

Note that y_i is identified with x_i for each i because the feedback line exists.

Threshold θ	Number of State Vectors
$n-1 < \theta$	$_nC_0 = 1$
$n-1 < \theta < n-1$	$_nC_1 = n$
$\vdots$	$\vdots$
$n-i-1 < \theta < n-i$	$_nC_i$ ($max.\ _nC_{\lfloor n/2 \rfloor}$)
$\vdots$	$\vdots$
$0 < \theta < 1$	$_nC_{n-1} = n$

Table 1 Relation between Threshold and Number of Stable Vectors

Table 1 shows the relation between the threshold and the number of stable vectors in the $_nC_k$ stable circuit. Comparing the $_nC_k$ stable circuit with Hopfield neural network of Figure 6, the negative output of the threshold gate is only utilized for feedback. All $_nC_k$ stable circuits can be realized in the same circuit only by changing the threshold. This is a great advantage of the threshold logic, and we also use it mainly in the prototyping stage, but the Boolean gate in the implementation stage, because of device sensitivity (Ae *et al* 1974).

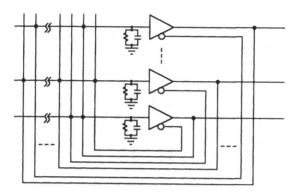

Figure 6 Hopfield neural network

For instance, the stable vector set of $_4C_k$ stable circuit is shown in Figure 7. The $_4C_1$ stable circuit is realized by Boolean gates as in Figure 8.

In the $_nC_k$ stable circuit, the Hamming distance of two stable vectors is at least 2 (as shown in Figure 7), and therefore, the trigger is required to change a stable vector to another. Note that each unstable vector can reach a stable vector within one cycle time of the host processor. Other multi-stable circuits (even with non-threshold logic gates) are also applicable for the building block method, if their characteristics have been known. The multi-stable circuit can be regarded as an associative memory (Kohonen 1977).

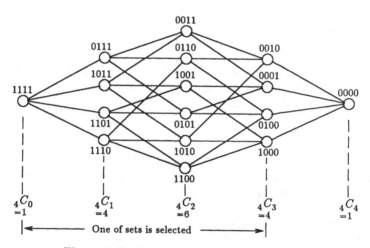

Figure 7 Stable vector sets of $_4C_k$ stable circuit

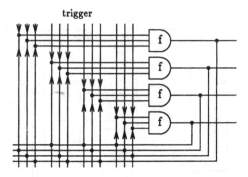

Figure 8 $_4C_1$ stable circuit realized by Boolean gates

ACCELERATOR FOR COMBINATORIAL PROBLEMS

As a typical building block architecture, we propose the orthogonal neurocomputer architecture (Ae and Aibara 1988a) which realizes the computation by the building block method for the combinatorial problem represented by a matrix form.

The matrix form is shown as in Figure 9, and Figure 10 describes an example, namely, a task scheduling problem, where the available number of processors at each time and the length of each task are denoted at the top and at the right-hand side, respectively. This number corresponds to the number of 1's in each row or column, but the 1's in each row must be consecutive in this example. (The latter condition is an additional requirement.) In Figure 10 an initial state is described, but the initial state is sometimes not necessarily required in the neural computation. The matrix-formed problem is implemented in the orthogonal neurocomputer architecture (in short, ONA) as follows:

Row condition $(i=1, \cdots ,p)$
   i-th row: k_i-out-of-n_i bits are 1, where the n_i $(n_i \leq p)$ positions may be designated and k_i may have additional requirements (e.g., its position must be connective), and
Column condition $(j=1, \cdots ,q)$
   j-th column: ibid

Each row or column corresponds to a fundamental block, i.e. the $_nC_k$ stable circuit, and therefore, the ONA of size $p \times q$ consists of $p+q$ $_nC_k$ stable circuits (as shown in Figure 11) and therefore consists of $2pq$ threshold logic gates and $O(pq^2+qp^2)$ interconnections.

In order to execute the neural computation in the ONA, the schema using primitives in Figure 4 must be prepared. The controller realizes the schema by adjusting the interconnection among rows and columns. The ONA is described as in Figure 12, and one step of the schema is mapped there by applying either of two operations in Figure 13. The step required for the neural computation is always within linear time (usually, within a constant time).

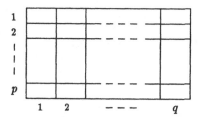

Figure 9 Matrix form for description of combinatorial problem

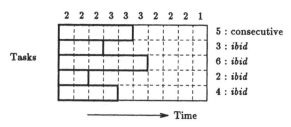

Figure 10 Task scheduling problem

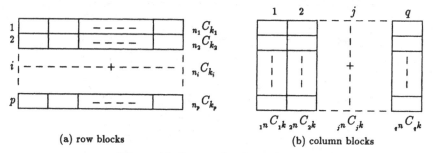

Figure 11 Row and column blocks in ONA

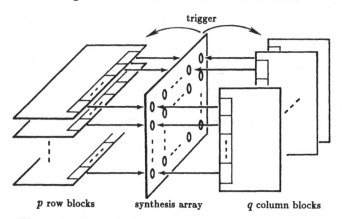

Figure 12 ONA: Orthogonal Neurocomputer Architecture

An example of Figure 13 using the non-feedback operations is realized as in Figure 14 which can solve a problem of the matrix form of size 3×2, within a constant time. This example is already fabricated as a prototype using discrete devices.

In order to realize the fundamental neural network, a circuit using PLAs or ROM/RAM is appropriate, especially, in the implementation stage.

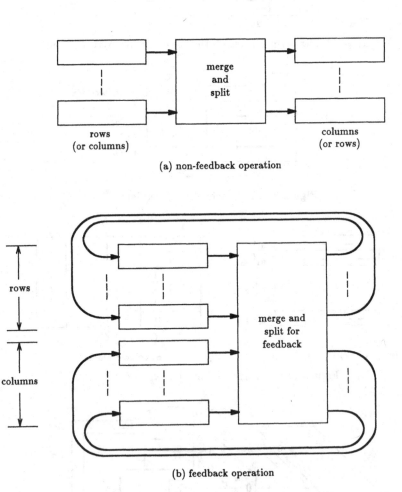

(a) non-feedback operation

(b) feedback operation

Figure 13 Two kinds of operations in ONA

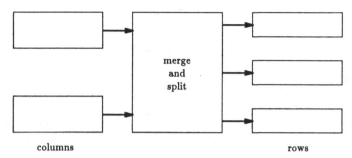

columns

rows

Figure 14 A simple example

3-D VLSI REALIZATION

Using 3-D VLSI technology, the fundamental neural network could be realized as in Figure 15, which consists of two layers, i.e., the device layer and the interconnection layer. The interconnection layer can be fixed, because the neural element of the device layer is programmable and realizes any Boolean function. Fifty neural elements are assumed to be realizable within a chip using nMOS technology, and therefore, one hundred chips compute the matrix-formed combinatorial problem of a size 50×50. We are now constructing a prototype of 8×16 by using conventional ICs, to evaluate the ONA for combinatorial problems.

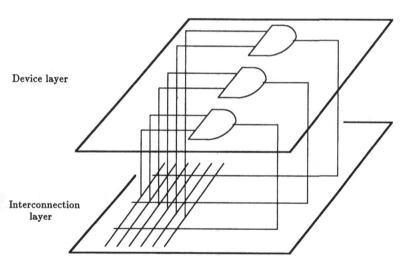

Device layer

Interconnection layer

Figure 15 3-D VLSI realization of fundamental Neural Network

REFERENCES

Ae, T., Nagami, H. and Yoshida, N., "On multi-stable transistor circuits using threshold logic operation" *International Journal of Electronics*, 36, 6, pp.849-856, 1974.

Ae, T. and Yoshida, N., "Multistable circuits and their coding," *Trans. IECE of Japan*, 58-D, 3, pp.129-134, 1975, in Japanese.

Ae, T. and Fujita, S., "A real-time image processing using optically-connected 3-D VLSI architecture," *Proc. SPIE Vol.804 Advances in Image Processing*, pp.70-77, 1987.

Ae, T. and Aibara, R., "Massive parallelism of a neurocomputer and its realization," *National Convention Record 1988, IEICE of Japan*, SD-4-3, Mar. 1988a in Japanese.

Ae, T. and Aibara, R., (to appear) Orthogonal Neurocomputer Architecture for a Class of Combinatorial Problems, 1988b.

Fujita, S., Aibara, R., Yamashita, M. and Ae, T., "A template matching algorithm using optically-connected 3-D VLSI architecture," *Proc. 14th International Symposium on Computer Architecture*, pp.64-70, Pittsburgh, June 1987.

Fujita, S., Aibara, R. and Ae, T., "A real-time production system architecture using 3-D VLSI technology," *Proc. 5th International Workshop on Database Machines*, pp.369-380, Karuizawa, Oct. 1987.

Hopfield, J. J. and Tank, D. W., " 'Neural' computation of decisions in optimization problems," *Biol. Cybern.*, 52, pp.141-152, 1985.

Kohonen, T., *Associative Memory—A System-Theoretical Approach*, Springer-Verlag, 1977.

6.4 SHIFT INVARIANT ASSOCIATIVE MEMORY

Donald Prados and Subhash Kak

INTRODUCTION

For AI applications related to pattern and image recognition, an associative memory should be able to recognize patterns or parts of patterns that are shifted spatially with respect to the ones that are stored. Furthermore, an associative memory should be able to recognize a pattern having a missing part despite the remaining parts being bunched together, as in a word having a missing letter. For the case of linearly shifted patterns, one can implement shift invariance by means of a pre-processor that performs a shift-invariant transformation such as a Fourier transformation. The associative memory is then used for storing the transformation coefficients. The recent speech recognition system of Kohonen (1988), for example, uses an FFT pre-processor. Alternatively, one may use a shift invariant neural network that accepts patterns directly rather than their transforms. With such a network, it may be easier to perform direct non-linear operations that allow recognition of patterns with missing or extra parts. Shift-invariant associative memory (SIAM) using neural networks has been described earlier by Maxwell *et al* (1986), Widrow and Winter (1988), and Prados (1988). In this contribution we focus on questions related to capacity and connection complexity in a hardware implementation.

CONVENTIONAL NEURAL NETWORK MODELS

Current neural networks require an astronomically large number of connections between neurons to perform as an associative memory. A neural network based on the Hopfield model, for example, would require over 16 million weights to store a 64 x 64 binary image. One's goals for a neural-network-based associative-memory model should thus include keeping the number of connections to a minimum and providing high capacity for a given number of neurons. In addition, one would like ease of initializing the connection weights as well as ease of modifying them should it be necessary.

The Hopfield model

Hopfield was instrumental in the recent renewal of interest in neural networks. In Hopfield (1982), he showed that neural networks can be seen as seeking minima in energy landscapes. If the minima are the stored memories, then it becomes clear how a neural network will serve as an associative or content-addressable memory. A subpart of sufficient size will yield the entire memory.

The Hopfield model uses the following equation to calculate the output of the ith neuron:

$$V_i = \text{sgn}\left\{ \sum_j T_{ij}V_j - I_i \right\} \tag{1}$$

where V_i is the output of the ith neuron, I_i is the threshold of the ith neuron, and T_{ij} is the synaptic weight corresponding to the influence of the jth neuron upon the ith neuron.

Given N neurons, i = 0,...,N-1, each of the N neurons has inputs from each of the other N-1 neurons and a synaptic weight corresponding to that connection. There is no direct feedback ($T_{ii} = 0$, for all i). The state of the neural network is simply the vector made up of the outputs of the N neurons. The model is viewed as being synchronous if each V_i is calculated simultaneously (each next state based on the previous state). It is viewed as being asynchronous if the calculation of each neuron takes into account all previous calculations (one neuron's output is calculated at a time).

To 'input' a pattern, one sets the outputs of the neurons to the binary sequence of the pattern. If one inputs a pattern that is not a stable state, the outputs continually change based on the above equation until a stable state is reached. To store a pattern, one must make that pattern a stable state by appropriately modifying the synaptic connection matrix T.

Hopfield utilizes an information storage algorithm inspired by Hebb. If one wishes to store the set of states V^s, s = 1,...,m, one uses the storage algorithm

$$T_{ij} = \sum_s V^s_i V^s_j \tag{2}$$

but with $T_{ii} = 0$. Notice that this equation will produce a symmetric connection matrix with $T_{ij} = T_{ji}$. The number of nonredundant connections is thus N(N-1)/2 (or N choose 2).

Example 1: Let the stored memories be $V^1 = (+ + + +)$, $V^2 = (+ + + -)$, and $V^3 = (- - + +)$. Using Hebbian learning, the connection matrix is calculated as follows:

$$T = \begin{bmatrix} 0 & 3 & 1 & -1 \\ 3 & 0 & 1 & -1 \\ 1 & 1 & 0 & 1 \\ -1 & -1 & 1 & 0 \end{bmatrix} \tag{3}$$

A simple check will reveal that this T does not actually store V^1 since it is too close to V^2. Furthermore, (- - - +) is a spurious stable stable state that arises spontaneously. In fact it is the complement of V^2. Spurious states are generally linear combinations of stored states.

The capacity of the Hopfield network depends on the manner in which it has been trained. Without unlearning of the spurious states, the capacity is roughly $N/\log N$; with unlearning of spurious states, it increases to N (Potter 1987, quoted by Bachmann 1987). The storage characteristics can be improved by means of asynchronous control (Stinson and Kak 1988). Note further that by allowing self-feedback, a network could store all possible patterns as stable states (T = I, the identity matrix); of course, that would make the network useless.

Modifications to the Hopfield model

There are several major problems with the Hopfield model. Disadvantages include the one-shot form of learning as opposed to incremental learning, a very limited degree of control over the basins of attraction, the fact that the patterns must be linearly separable if they are to be distingushed using the Hebbian storage matrix, and the inability to deal with spatially shifted patterns.

The most common model that utilizes incremental learning is the Widrow-Hoff algorithm. This algorithm significantly increases the number of patterns that can be stored compared to the Hebbian algorithm. If pattern V^s is not successfully stored, then, for each bit i of V^s that changes when V^s is applied to the network, modify each T_{ij} by

$$\Delta T_{ij} = c \, V^s_i \, V^s_j \tag{4}$$

where c is a learning constant.

A spurious state V^s may be unlearned by subtracting $V^s_i V^s_j$ from each T_{ij}. A learning constant similar to that in Equation 4 may be used to control the degree of unlearning. In models that are sufficiently large, one would not have the computational capability to determine these spurious states, so unlearning may not be a viable part of the training proceedure.

Higher order terms

The Hopfield neuron discussed above can be classified as a perceptron of order one (Minsky and Papert 1969). The perceptron, as defined by Minsky and Papert, is a device capable of computing all predicates which are linear in some given set of predicates. It has order n if no member of the set of predicates depends on more than n points. Since the Hopfield neuron simply calculates a linear combination of the inputs, it can implement only linear functions. The classic example of a nonlinear function is the exclusive-or function. If a two-input first-order perceptron is off for 00 (when both inputs are off), on for 01, and on for 10, it will be on for 11 due to its linear nature; thus, it cannot perform the exclusive-or function.

By increasing the order to two, the perceptron is capable of performing the exclusive-or function. The output of neuron i can be determined by the following equation:

$$V_i = sgn\left\{\sum_j T_{ij}V_j + \sum_j\sum_k T_{ijk}V_jV_k\right\}$$ (5)

Such a neural network can be trained by using the following Hebbian-type of equation in conjunction with Equation 2:

$$T_{ijk} = \sum_s V^s_i V^s_j V^s_k$$ (6)

Notice   that   this   equation   will   produce   the   symmetries $T_{ijk} = T_{ikj} = T_{jik} = T_{jki} = T_{kij} = T_{kji}$. With such symmetry and no direct self-feedback of neurons $(T_{ii} = T_{iji} = 0)$, the second order connection matrix requires $N(N-1)(N-2)/6$ (or N choose 3) nonredundant weights.

Using this symmetry and the fact that $V_j V_k = V_k V_j$, equation 5 can be modified to

$$V_i = sgn\left\{\sum_{j<i} [T_{ij}V_j + \sum_{k<j} T_{ijk}V_jV_k]\right.$$

$$\left. + \sum_{j>i} [T_{ji}V_j + \sum_{k<j,l>i} T_{jki}V_jV_k + \sum_{k<j,k<i} T_{jik}V_jV_k]\right\}$$ (7)

where only those T_{ijk} with $k<j$ and $j<i$ are used.

The Widrow-Hoff learning rule can easily be extended to second order terms (Maxwell *et al* 1986). The second order weight matrix can be updated by

$$\Delta T_{ijk} = c V^s_i V^s_j V^s_k.$$ (8)

The general higher-order correlation model has the characteristic equation

$$V_i = sgn\left\{\sum_j T_{ij}V_j + \sum_j\sum_k T_{ijk}V_jV_k + \sum_j\sum_k\sum_l T_{ijkl}V_jV_kV_l + \cdots\right\}.$$ (9)

Note that the number of connections increases dramatically with an increase in the order of correlations used. For the 2-dimensional (d=2, first order) Hopfield-type connection matrix $(T_{ij} = T_{ji})$, there are "N choose 2" nonredundant connections. In general, if one includes only the nonredundant weights, there will be "N choose d" connections. The symmetry reduces the number of connections by a factor of 'd factorial' from the case in which there is no symmetry. The number of connections for a matrix of dimension d + 1 is (N - d)/(d + 1) times greater than the number for a matrix of dimension d. (If d is greater than N/2, however, this factor will be less than 1, and there will actually be fewer connections for a connection matrix of dimension d + 1 than for a matrix of dimension d). For example, if N = 100, increasing the dimension of the weight matrix from 2 to 3 results in an increase in the number of non-redundant weights from 4950 to 161,700 (approximately 33 times as many weights). Whereas Hopfield was able to store approximately fifteen patterns successfully using his Hebbian model with N = 100; Chen *et al* (1986) were able to store approximately 500 patterns using a 3-dimensional connection matrix model with N = 100 (approximately 33 times more patterns). Maxwell *et al* (1988) have found that, for a given d, the storage capacity is proportional to N_w/N, where N_w is the number of non-redundant connection weights.

As noted earlier, if one allows self-feedback of neurons, one can store any set of patterns by allowing the connection matrix to approach the identity matrix in form. If self-feedback is not allowed, use of higher order terms leads to the following theorems concerning the limit on the number of patterns one can store (Prados 1988).

Theorem 1. For a set of patterns to be successfully stored in a neural network without self-feedback, all pairs of patterns in the network must differ by more than one bit.

Proof. If a neural network does not have self-feedback of neurons, the calculation of the next state of a neuron will depend on the outputs of all the other neurons and the appropriate connection weights but not on the state of the neuron itself. Assume two patterns differ in only one bit. Whether the state of this neuron is +1 or -1, the calculation of its next state will be identical. Thus if two patterns differ in only one bit, the next state of the pattern will be the same for both patterns. If one of the two patterns is a stable state, the other will have as a next state that same stable state. Note that this theorem implies that no more than 2^{N-1} patterns can be simultaneously stored in a network without self-feedback.

Theorem 2. A neural network without self-feedback can store 2^{N-1} patterns.

Proof. The set of all even (or all odd) parity patterns can be stored in a neural network. The output of each neuron can be determined by knowledge of each of the other N-1 neurons. If N is even, the next state of each neuron can be obtained by taking the product of the other N-1 neurons. If N is odd, the next state of each neuron can be obtained by multiplying the product of the other N-1 neurons by -1. Notice that, if the outputs are limited to +1 and -1, multiplying them is equivalent to performong the exclusive-nor operation.

The following example shows how 2^{N-1} patterns can be stored using higher order terms.

Example 2. Let N = 4 and use a 4-dimensional connection matrix for calculation of the next state:

$$V_i = \text{sgn} \left\{ \sum_j \sum_k \sum_l T_{ijkl} \, V_j \, V_k \, V_l \right\} \tag{10}$$

Let $T_{1234} = T_{2134} = T_{3124} = T_{4123} = 1$ and all other $T_{ijkl} = 0$. Such a choice will success-fully store the eight even parity patterns of length 4. In general, a connection matrix of dimension N that includes only N nonzero weights is needed to store all even (or all odd) parity patterns of length N, one weight for each neuron.

SHIFT INVARIANT MODEL

The conventional neural network model needs to be modified only slightly to provide shift invariant pattern recognition. Imposing shift invariance on the output of a neural network is equivalent to imposing the constraint that the connection matrix depend only on relative coordinates, not absolute coordinates. Suppose, for example, that one wishes to store the following two patterns in a neural network memory: V^1 = (- - - + + +) and V^2 = (- + - + - +). Associated with each of the 6 neurons is a set of weights corresponding to the effect of each of the other neurons upon it. For example, neuron 2 (the third neuron) should have a positive weight reflecting an excitatory effect of neuron 0 upon it and negative weights reflecting inhibitory effects of neurons 3 and 5 upon it. The Hopfield model using the Hebbian storage algorithm would assign weights of $T_{20} = +2$ and $T_{23} = T_{25} = -2$. A shift invariant neural network should assign weights in accordance with the relative distances between neurons. The weight T_2, for instance, can be calculated as the average of T_{20}, T_{31}, T_{42}, and T_{53}.

The Hopfield equation, Equation 1, can be modified to obtain

$$V_i = \text{sgn} \left\{ \sum_j T_j V_{i-j} \right\} \tag{11}$$

where V_i is the output of neuron i and T_j represents the connection strength between neuron i and neuron i - j. T_j depends only on the distance between neurons.

The matrix T can be calculated using the Hebbian-type calculation:

$$T_j = \sum_s \sum_i V^s_i \, V^s_{i-j} \tag{12}$$

A Widrow-Hoff-type incremental-learning rule can be used to modify the connection matrix to store additional patterns. To store pattern V^s, input V^s to the neural network. If each bit of the output V (next state of V^s) is equal to the corresponding bit of V^s, no changes are necessary. For each bit that changes, modify T as follows:

Case 1. $V^s_i = -1$ but $V_i = 1$: Decrement each term in Equation 11. If $V^s_{i-j} = 1$, decrement T_j. If $V^s_{i-j} = -1$, increment T_j.

Case 2. $V^s_i = 1$ but $V_i = -1$: Increment each term in Equation 11. If $V^s_{i-j} = 1$, increment T_j; otherwise, decrement T_j.

In general, this algorithm can be written as:

$$\text{if } V_i \neq V^s_i, \text{ then, for each } j, \Delta T_j = V^s_i V^s_{i-j} \tag{13}$$

An alternative form of this equation is

$$\Delta T_j = c \left[V^s_i - \text{sgn} \left\{ \sum_j T_j V^s_{i-j} \right\} \right] \cdot V^s_{i-j} \tag{14}$$

where c is a learning constant.

To increase the number of patterns that can be stored, higher order terms can be used:

$$V_i = \text{sgn} \left\{ \sum_j T^{1d}_j V_{i-j} + \sum_j \sum_k T^{2d}_{jk} V_{i-j} V_{i-k} + \sum_j \sum_k \sum_l T^{3d}_{jkl} V_{i-j} V_{i-k} V_{i-l} + \cdots \right\} \tag{15}$$

The Widrow-Hoff-type equations for second and third order terms are:

$$\Delta T^{2d}_{jk} = V^s_i V^s_{i-j} V^s_{i-k} \tag{16}$$

$$\Delta T^{3d}_{jkl} = V^s_i V^s_{i-j} V^s_{i-k} V^s_{i-l} \tag{17}$$

Notice that use of these equations will not destroy the symmetry of the connection matrices. A d-dimensional connection matrix will require only "N-1 choose d" non-redundant connection weights whether or not the Widrow-Hoff algorithm is used. Also notice that a shift-invariant model with a 2-d connection matrix can perform the exclusive order function (it utilizes second order correlations).

If a pattern is stored using either the Hebbian (Equation 12) or Widrow-Hoff (Equation 13) type of calculation and all subtractions in the indices are performed modulo N, all cyclic shifts of the stored pattern will also be stable states. As in the Hopfield model, no two states can simultaneously be stored that differ by only one bit. Thus, if a set of patterns is to be stored, each pattern must differ in more than one bit from all other patterns in the set. Under such constraints, one should be able to successfully store any set of patterns by sufficiently increasing the order of of the correlations used.

The storage of 2^{N-1} patterns using a higher order correlation matrix requires a connection matrix of dimension N with only one nonzero weight. For example, if N = 4, use of a 3-dimensional connection matrix where $T_{123} = +1$ and all other $T_{ijk} = 0$

allows the storage of all even parity patterns.

As the number of patterns to be stored approaches the capacity of the neural network model, the ability to correct errors decreases. If one has a certain amount of error correction as a goal, the theory of cyclic codes can be used to obtain a limit on the number of patterns one should attempt to store.

Consider a conventional (N,L) cyclic code. This model chooses an appropriate H-matrix generated by a polynomial h(D) of degree L which divides $D^N - 1$ over an appropriate field. To make comparisons with the binary neurons of neural networks, this field is taken to be GF(2). For a given pattern V to be a stable state of this model,

$$V H = 0. \tag{18}$$

The minimum Hamming distance between any two stable states is given by 2r + 1 where r is the solution to

$$\sum_{i=0}^{r} \binom{N}{i} \leq 2^{N-L}. \tag{19}$$

Such a model will detect and correct all errors of r bits or less. For a shift-invariant neural network to have an error correcting ability of r bits, the minimum Hamming distance between any two stable states must be greater or equal to 2r + 1. Since neural networks are not guaranteed to produce the stable state closest in Hamming distance to an input pattern, one should avoid coming too close to this limit. Equation 19 can also be used as a limit on the number of patterns one attempts to store given a length N and minimum error correcting ability r by solving for L. For example if N = 7 and one requires error correction of 1 bit, one should not attempt to store more than 4 patterns. Keep in mind that one would actually be storing the 4 patterns plus all of their cyclic shifts.

CONCLUSION

We have seen that the Hopfield neural-network model requires "N choose d" nonredundant connections. With the SIAM model there are "N-1 choose d" nonredundant connection weights. This is still a very large number for large N. For a 64x64 binary image, one would still need over 8 million nonredundant connections--not much less than required for the Hopfield model.

By limiting the connections to a neighborhood surrounding the neuron, this number can be reduced significantly. For example, if each neuron is connected to only those neurons within a 15x15 grid surrounding it, only about 25,000 connections would be required (for a first order network). A shift invariant neural network of this kind could be used as a low-level classsifier as part of a hierarchical system. Objects smaller than the neighborhood could be stored in the connection matrix. These small objects could then be recognized by the low-level classifier and application of a higher-level classifier could follow.

Note that, although there are estimated to be about 100 trillion synapses in the human brain, on the average, each of the more than 10 billion neurons in the nervous system has 100 inputs converging on it while it in turn diverges to 100 other neurons (Ganong 1979). Also, when one observes an image, for example a written page, one does not take in the entire image at once but focuses on only a small area of the image at a time. By limiting the connectivity of a neural network as described above, relatively small objects could be stored and could be recognized regardless of their location in the image without using an astronomically large number of connection weights. For a neural-network-based associative memory to be feasible such a reduction in the number of connection weights is essential.

REFERENCES

Bachmann, C. M., Cooper, L. N., Dembo, A., Zeitouni, O., "A Relaxation Model for Memory with High Density Storage," *Proc. National Academy of Science, USA,* 1987.

Chen, H. H., Lee, Y. C., Sun, G. Z., Lee, H. Y., Maxwell, T., and Giles, C. L., "High Order Correlation Model for Associative Memory," in *Proc. AIP Conference on Neural Networks for Computing,* Snowbird, UT, pp.86-99, 1986.

Ganong, W. F., *Review of Medical Physiology.* Los Altos, CA: Lange Medical Publ., 1979.

Hopfield, J. J., "Neural Networks and Physical Systems with Emergent Collective Computational Abilities," *Proc. National Academy of Science , USA,* vol. 79, pp. 2554-2558, 1982.

Kohonen, T., "The Neural Phonetic Typewriter," *Computer,* vol. 21, no. 3, pp. 11-22, 1988.

Maxwell, T., Giles, C. L., Lee, Y. C., and Chen, H. H., "Nonlinear Dynamics of Artificial Neural Systems," in *Proc. AIP Conference on Neural Networks for Computing,* Snowbird, UT, pp. 299-304, 1986.

Maxwell, T., Giles, C. L., and Lee, Y. C., "Transformation Invariance Using High Order Correlations in Neural Net Architectures," Plasma Preprint UMLPF #88-125, Univ. of Maryland, 1988.

Minsky, M. L., and Papert, S., *Perceptrons: An Introduction to Computational Geometry.* Cambridge, MA: MIT Press, 1969.

Prados. D., "The Capacity of a Neural Network," *Electronics Letters,* vol. 24, no. 8, pp. 454-454, 1988.

Stinson, M., and Kak, S. C., "Asynchronous Controller to Improve the Convergence of Neural Nets," in *Proc. ACM Southeast Conference,* Mobile, AL, pp. 410-413, April, 1988.

Widrow, W., and Winter, R., "Neural Nets for Adaptive Filtering and Adaptive Pattern Recognition," *Computer,* vol. 21, no. 3, pp. 25-39, 1988.

Chapter 7

DIGITAL AND ANALOG VLSI NEURAL NETWORKS

Most of the research on neural network has concentrated on theoretical studies and simulations on uni–processor architectures. Simulations are, however, slow; therefore it is essential to have hardware implementations of these networks which are highly parallel.

Although today's technology allows the integration of a large number of simple processors on a single chip, the high interconnectivity required by the neural networks is one of the major challenges. Building a VLSI neural network circuit, a researcher may face many alternatives that are, in general, extremely difficult to evaluate. For example, decisions have to be made on the weight accuracy, the storage devices, the multiplication and addition circuits, the communication scheme, the threshold function implementation, as well as choosing whether to implement a high or a limited interconnectivity, and whether to implement the learning algorithm on- or off–chip. A good design must take into account not only what is desired of a neural network system but also what is achievable in VLSI technology. In this chapter several implementations of neural networks are presented; each one has a different approach to exploit the VLSI capabilities and overcome the limitations imposed by the technology.

One of the most difficult choices that a designer must make is between a digital, an analog or a hybrid implementation. Each of these styles have advantages and drawbacks.

DIGITAL IMPLEMENTATIONS

A fully digital implementation of neural networks has some advantages: easier programmability, higher precision, noise–free, highly reliable storage devices, and technology independence. The main drawback is that arithmetic circuits are extremely large in comparison with analog devices; the size of the system will be dominated by these circuits.

A novel architecture that is an array of n^2 synaptic operators for n neurons is presented by Butler *et al* §7.1. Bit–serial and reduced arithmetic are adopted not only to minimize the interconnection requirements but also to reduce silicon area. The biological neural activation function, a sigmoid, is approximated by a five–state function. A 3×9 synaptic array has been designed; to achieve a suitable size several chips are included on a board and a *paging* scheme is also used. Another digital implementation is proposed by Weinfeld (see §6.2).

ANALOG IMPLEMENTATIONS

Using analog computation has some advantages: a smaller circuit size , higher speed and possibly a lower power consumption. However, large integrated analog circuits may be difficult to design; problems such as weight storage, precision, and programmability must be addressed in order to implement analog networks.

An architecture that has only nMOS transistors as active synapses and differential amplifiers as neurons is presented by Verleysen *et al* §7.2. A synapse acts as a programmable current sink or source and provides three discrete values. Each neuron has two inputs that form the addition of currents of the synapses connected to it. A chip with 14 neurons and 196 synapses is presented.

Today's VLSI technology cannot support the extremely high interconnectivity found in biological neural networks. Akers *et al* §7.3 have designed and implemented a limited–interconnect architecture using a multi–layer approach. An analog neural cell computes the weights and sum of products. The weights are stored dynamically on the gates of transistors and loaded by means of multiplexed lines. A 512–element feedforward array is described in the paper.

Rückert and Goser §7.4 present two implementations of adaptive connection element. The first is an analog circuit that has floating gate transistors to store information. The second implementation is based on CCD loops that hold the data.

A system that mimics biological neural network is introduced by Murray *et al* §7.5. The neuron circuit is fed with excitatory and inhibitory inputs which are pulse streams; the ouput is another pulse stream whose frequency is a function of the inputs. Two circuits are presented to store the weights: one is a digital memory and the second is an analog storage device.

7.1 VLSI BIT–SERIAL NEURAL NETWORKS

Zoe Butler, Alan Murray and Anthony Smith

INTRODUCTION

A synthetic neural network can be viewed as a large parallel array of n^2 synaptic operators (for n neurons) that is able to model some of the brain's characteristics. The VLSI neural network described, functions with bit-serial, two's complement arithmetic and uses a single phase clocking technique operating at a minimum of 20 MHz (McGregor *et al* 1987).

A synthetic neuron is a state machine that is either "on" or "off", assuming intermediate states as it switches smoothly between these extremes. A synapse weights the signal from a transmitting neuron such that it is more or less excitatory or inhibitory to the receiving neuron. The total level of activation of a neuron is represented by its *activity*, x_i. This is related to the state of the receiving neuron by an activation function, f, that describes its response to a change in activation. Biologically, this function is sigmoidal, but in our synthetic network it is simplified so that $V_i = 1$ when x_i is large and -1 when x_i is small, with 3 states in between. The interneural *synaptic weights*, T_{ij}, are the contributions from other neurons, that are weighted by the receiving neuron. Therefore, the state of neuron i in an n-neuron array is given by

$$V_i = f(x_i) = f(\sum_{j=0}^{n-1} T_{ij} V_j + I_i) \tag{1}$$

Synaptic weights may be positive (excitatory) or negative (inhibitory) and any neuron may tend to turn any other neuron "on" or "off" respectively. I_i is a direct input that may be arbitrarily strong to force some value on V_i. The synaptic weights, determine the stable states and represent the information learned by the network. *Learning* is therefore, a controlled modification of the $\{T_{ij}\}$ to adjust the stable states. Recall or computation is performed as the network moves around the n - dimensional space defined by the neural states V_j, with the $\{T_{ij}\}$ constant.

The neural architecture is based on Equation 1. It involves n^2 digital multiplications and summations in an array of n totally interconnected neurons. This is relatively straightforward in a network with fixed functionality and modest n. However, if the network is to be able to learn patterns, the synaptic weights must be programmable, thus making it more complicated.

NETWORK COMPUTATION AND DESIGN

An advantage of bit-serial arithmetic in a neural network is that it minimises the interconnect requirement by eliminating multi-wire busses. Pipelining makes optimal use of the high bit-rates possible in serial systems allowing good communication within and between VLSI chips. The primary advantage of using digital CMOS circuitry is that on-chip digital memory design is easier to implement than on any analogue counterpart and can be incorporated for the programming and storage of the synaptic weights. Design techniques are advanced, automated and well understood, and noise immunity and computational speed can be high.

Architecture

The neural architecture in Figure 1 shows a single network of n totally interconnected neurons. A neuron is represented by a circle, with its column of n synapses (shown by squares) communicating with all other neurons in the array. Each synaptic operator *adds* the weighted contributions from other neurons down the column. When the total summation reaches the foot of the column, the neuron thresholds it according to the 5-state activation function shown in Figure 2. The new state of the neuron is then signalled back to the array. The state signals are connected to a n-bit bus running across the synaptic array, with a connection to a synaptic operator in every column. Therefore, the two functions of a synaptic operator are to multiply the signalling neuron state V_j by the synaptic weight T_{ij} and to add the product to the running total of activity. For example, in Figure 1, neuron 3 signals its state V_3, to neuron 1 along the dark path shown, and the product $T_{1,3}V_3$ is added to the running total in column 1.

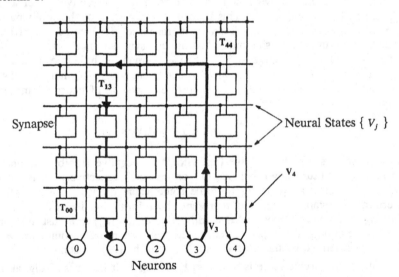

Figure 1 Generic Architecture for a totally interconnected n-neuron network.

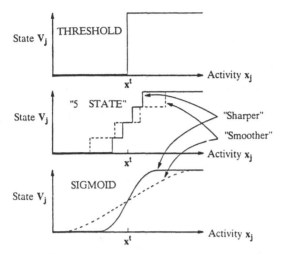

Figure 2 The 5-state, sigmoid and 2-state activation functions.

Reduced Arithmetic

Full digital multiplication can be expensive in silicon area, but the 5-state activation function allows reduced arithmetic to be used. Hence, multiplication of a synaptic weight by $V_j = 0.5$ simply requires the synaptic weight to be right-shifted by 1 bit. Likewise, multiplication by 0.25 involves two right-shifts of $\{ T_{ij} \}$, and multiplication by 0 is easy. A negative (inhibitory) synapse is straightforward, as a switchable adder/subtractor 25% transistors than are required for an adder. Therefore, 5 neural state synapse can be obtained from circuitry 50% larger than that required for 2 states (Hopfield 1982). The neural state bus expands from a 1 bit to a 3 bit representation, where the 3 control bits are add/subtract?, shift? and multiply by zero?

Details of a synaptic operator are given in Figure 3. Each operator has an 8 bit shift register memory holding its synaptic weight. During computation, the synaptic weight cycles round the register while the neural state is signalled on the 3 bit bus running horizontally above each synaptic row. A complete synapse computation requires two complete shift register cycles (16 clock cycles). During, the first cycle the synaptic weight is multiplied by the neural state and during the second, the most significant bit of the resultant $T_{ij} V_j$ is sign-extended for the remainder of the shift register cycle. This allows a maximum 8 bit word growth in the running summation. The least significant bit of each neuron's running summation is indicated by a least significant bit signal running down the synaptic column.

The final 16 bit summation at the foot of the column is thresholded according to its activation function. As the neuron activity x_j, increases through threshold value x, (Figure 2), the ideal activation represents a smooth switch of neural state from -1 to +1. The 5-state "staircase" function gives a better approximation to this than the 2-state threshold function. Control of the sharpness of this transition can "tune" the neural dynamics for learning and computation. The control parameter is

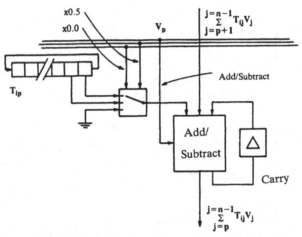

Figure 3 Synaptic Operator with a 5-state activation function.

referred to as temperature by analogy to statistical functions with this form. Higher temperatures give the staircase and sigmoid a lower gradient.

LEARNING AND RECALL OF THE ACTIVATION FUNCTIONS

Software simulations of learning and recall capabilities of the 5-state model were compared with those of the 2-state and sigmoid activation functions at varying temperatures with a restricted dynamic range for the synaptic weights. A 64 node network in each simulation attempted to learn 32 patterns using the delta rule algorithm (Rumelhart 1986). The learned patterns were then corrupted with 12.5% noise. Results showed that the 5-state activation function recalled the weight sets with considerably greater noise immunity than the 2-state activation function. The sigmoid activation was still superior to the 5-state in recalling the corrupted patterns, but the discrepancy was noticeably less than that between the 5-state and the 2-state activation functions. The best method to deal with weight saturation during learning was to permit any weight outside the dynamic range to be set to its maximum value. A full discussion of these results can be found in Murray *et al*, 1987.

HARDWARE NEURAL BOARD

A 5-state synaptic operator array is being fabricated in 3µm CMOS technology. Full custom layout allowed a 12 x 9 synaptic array in a 64 pin package and Figure 4 shows part of the design. Several chips, therefore, need to be wired together with memory ICs and control circuitry to achieve a suitable size network for simulations.

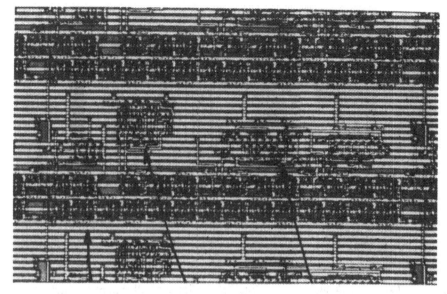

8 bit shift register   neural state tree   sum/carry tree

Figure 4 Silicon Layout of the Synaptic Array.

Neural Paging Architecture

A neural board has been designed with 4 synaptic chips wired together giving a 12 x 9 synaptic array. The small array will be used in a *paging* architecture to give a network of 256 neurons that will act as a *neural accelerator* to a host computer. The *paging* architecture can be thought of as a "moving patch", where the small array or patch will simulate a small number of synapses in a large array, and then pass onto the adjacent patch to repeat the computation until all 256 synapses have been simulated. This idea is shown in Figure 5. Each time the array is moved to represent another set of synapses, the weights for that patch must be loaded into it. The memory required for 256 neurons is 0.5 Mbits of static RAM. A RAM speed of 70ns will allow the weights to be loaded in 9ms. A larger number of neurons can be simulated by simply loading the extra synaptic weights into more memory.

The "patch" will move down the 1st set of 12 columns to compute the complete running activities. It will then compute the 2nd set, 3rd set etc., until each set has been computed. For each "patch" simulation in the array, the emerging partial running summations of the 12 *partial* column blocks, are synchronised to coincide with the top of the running summation of the new patch. This ensures that each column has a contribution (excitatory or inhibitory) from each synapse. As the total summations occur for each block, they are stored in an on-board static RAM as indicated in the board design in Figure 6.

When the total summation has been completed in each column, the neurons' activities are thresholded off-board according to the 5-state activation function. The new neural states are signalled back to the synaptic accelerator chips for the next

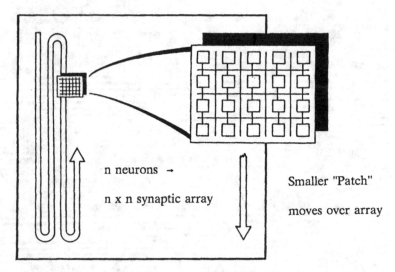

Figure 5 "Paging Architecture" of passing a small synaptic "patch" over a larger n x n synaptic array.

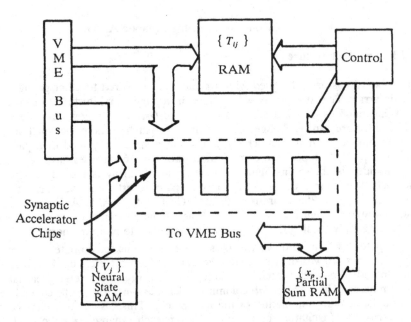

Figure 6 "Paging Architecture" for a Neural Network Board.

array computation. Once the states become stable, the synaptic weights are adjusted accordingly until learning is complete.

Control Circuitry

Microcode circuitry controls the RAM access and the synaptic accelerators. The flow diagram in Figure 7 shows the small control overhead required, along with the timing of all operations for a complete update of 256 neurons. The calculated update time for the board is 1ms giving 6 x 10⁷ operations/second. The number of synaptic accelerators determines the operating speed. A faster speed, or more neurons would require more accelerators. Hence, the design is versatile in that a specification for network size and speed can be met.

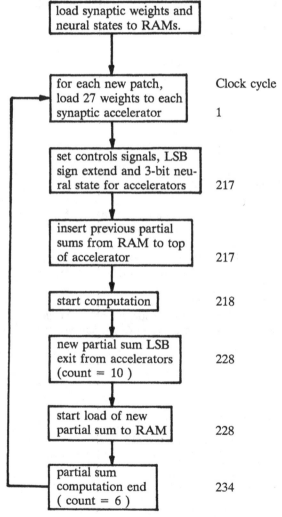

Figure 7 Flow Diagram of the Control Operation.

CONCLUSIONS

The design method has been given for the construction of a VLSI neural hardware accelerator and its implementation in a neural board. Bit-serial, reduced arithmetic improved the level of integration compared to more conventional digital, bit-parallel schemes. The restrictions on synaptic weight size and arithmetic precision by VLSI constraints have been examined and proved to be tolerable, using the associative memory problem as a test.

The digital design gives a good compromise between arithmetic accuracy and circuit complexity, but the level of integration is disappointingly low. This has been offset by the paging architecture of the neural board to enable the simulation of a large number of neurons. It is our belief that, while digital approaches are useful in the medium term, especially as hardware accelerators, analogue techniques represent a better option in 2-dimensional silicon.

The authors acknowledge the support of the Science and Engineering Research Council (UK) in the execution of this work.

REFERENCES

Hopfield, J. J., "Neural Networks with Emergent Collective Computational Abilities," *Proceedings of the National Academy of Science, USA*, vol. 79, pp. 2554-2558, 1982.

McGregor, M.S., Denyer, P.B. and Murray, A.F., "A Single - Phase Clocking Scheme for CMOS VLSI," *Advanced Research in VLSI: Proceedings of the 1987 Stanford Conference*, 1987.

Murray, A.F., Smith, A.V.W. and Butler, Z. F., "Bit-serial Neural Networks," *IEEE Conf. on Neural Infomation Processing Systems - Natural and Synthetic, Denver*, 1987.

Rumelhart, D.E., Hinton, G.E. and Williams, R.J., "Learning Internal Representations by Error Propagations", *Parallel Distributed Processing: Explorations in the Microstructure of Cognition*, vol. 1, pp. 318-362, 1986.

7.2 A NEW CMOS ARCHITECTURE FOR NEURAL NETWORKS

Michel Verleysen, Bruno Sirletti and Paul Jespers

INTRODUCTION

Conventional computer architectures and artificial neural networks have radically different structures and utilities. While computers are used to quickly perform complex but clearly defined tasks, neural networks are particularly adapted to solve more blurred problems like those encountered in perception, recognition and optimization. The hope is that neural networks might share some of the processing capabilities of the human brain. This is especially interesting in perception problems which are very long to compute by conventional computers because of the need for powerful sequential algorithms. The tradeoff between speed and efficiency implies the use of more powerful computers in order to run the complex algorithms needed to solve real-time perception problems. Nevertheless sequential solutions are not "natural" for this class of problems in which a large amount of simple but repetitive computation is needed.

In a human brain, about 10^{12} neurons form a three-dimensional, highly interconnected network. Each neuron can be connected up to 10^4 other neurons (Rumelhart and McClelland 1986); the huge computational power of the brain resides only in this highly parallel structure: although the propagation times between the neurons and the neuron's delay time are very long, a complex vision problem can be solved in less than 500 milliseconds. It seems therefore obvious that no complex algorithm is used inside the brain.

The idea which lead to the design of artificial neural networks is the emulation of the brain's structure in order to take advantage of its perception properties. In the brain, each neuron can be seen as a single processing element which performs a weighted sum of its inputs (the outputs of the other neurons). The neuron turns on if the result is greater than an internal threshold (Fahlman and Hinton 1987). The difficulty in the realization of artificial neural networks resides in performing this weighted sum efficiently .

In image processing, to obtain interesting results, the network must be large enough to perform recognition tasks on a sufficient number of pixels. Since the power of parallel solutions in general, and neural networks in particular, comes from their ability to solve complex problems quickly, the processing speed of the networks must also be high enough to meet real time requirements. The recent advancements in the VLSI technology make possible the implementation of large fast networks due to the increasing number of transistors which can be integrated on a single chip.

ARTIFICIAL NEURAL NETWORKS

An electronic neural network can be seen as an array of simple processing elements (*neurons*) connected through a coupling network (a single connection between two neurons is called *synapse*). Each pair of neurons can be connected by a synapse which is either excitatory (positive) or inhibitory (negative) (Figure 1). All the information contained in the network resides in the connection values and is thus distributed within the whole system. The neurons only perform a weighted sum of all their inputs and a comparison between this sum and a fixed threshold.

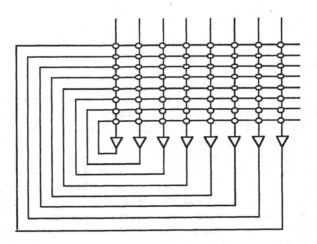

Figure 1 Fully interconnected neural network

In Hopfield's model, the neurons are first set to the input pattern values (Hopfield 1982). The neuron values are then reassigned by computing for each neuron the sum of all its inputs weighted by the synapse strengths; this processus is repeated until a stable state is reached.

The Hopfield's model allows the realization of a content-addressable memory by setting appropriate connection values. A simple rule inspired from biological models, called the Hebb's rule (Hebb 1949), is used to compute the different connection weights. It consists in increasing the strength between two neurons if they have the same state in the pattern to memorize, and in decreasing it if they don't. This simple rule can be easily formalised:

$$T_{ij} = \sum_p V_{ip} V_{jp}$$

where T_{ij} is the connection strength between neurons i and j

   V_{ip} is the value of neuron i (+1 or -1) in the pattern p to memorize.

Although this rule is very simple, it allows the realization of a content-addressable memory which can store up to 0.15N patterns of N bits (Lippmann 1987). More complex rules increase this 0.15N limit (Personnaz 1986).

VLSI NEURAL NETWORKS

Synapses using resistors

By analogy with the theoretical models of neural nets, the first implemented networks used resistors to realize the coupling matrix (Howard *et al* 1987). Each synapse contains a resistor whose value determines the strength of the connection. In order to allow positive and negative values, the resistors are connected either to the non-inverting or inverting output of the neuron.

Although this is the simplest and most powerful way to realize these networks, two disadvantages appear because of the use of resistors. First, the different connection strengths need resistors with different values which occupy various areas; this prevents the network from having an highly regular structure where each synapse would occupy the same area on the chip. Secondly, because the connection values must be set during the design of the chip, no programming and no learning are allowed. Since in most applications the patterns to be stored are not previously known, chips with fixed connection values do not cover the whole range of applications we can expect from neural networks.

These two restrictions strongly limit the use of resistors networks.

Synapses using current sources

A new design for VLSI neural networks has been proposed in recent papers (Graf and de Vegvar 1987, Tsividis 1987). In those circuits, each synapse is a programmable current source controlled by the output of the neuron to which the synapse is connected; following the sign of the connection and the one of the connected neuron, the synapse sources or sinks current to the input line of the second neuron to which the synapse is connected. In such implementations, the synapse possible values are generally limited to +1, 0 or -1. Indeed, if more values were allowed, more memory points would be necessary to store the connection weight; the area of one synapse would be larger, and less neurons could be implemented on a single chip. The current trend is to reduce the area of the synapse by using algorithms which show no performance decrease when they are used in a network with three-values synapses.

In the Hopfield's model, a XOR function must be realized in each synapse between the output of the connected neuron and the memorized weight. According to the result of this XOR function, a current is sourced by a P-type transistor or sunk by a N-type one (Figure 2). All the sourced and sunk currents are summed on the input line of the connected neuron. The logical function of the neuron is to detect if more or less synaptic currents are sourced than sunk. Since the only input of the neuron is the sum of all the synaptic currents, we must detect if this sum is greater or less than the threshold (0 in our case). A problem arises immediatly: the P-type and N-type current sources will never be exactly equal because of the mobility differences between the two types of charges. This mismatching is multiplied by the number of active synapses, and can quickly reach the value of one synaptic current; this limits considerably the number of neurons which can be connected together. Figure 3 shows the maximum number of neurons versus the ratio mismatching/single synaptic current. It shows that with the physical feasible values for this ratio (about 3-5 %) (Nicollian and Brews 1982), the number of neurons is strongly limited.

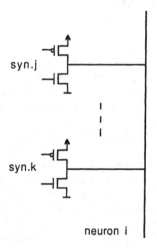

Figure 2 P and N current sources

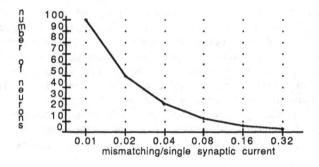

Figure 3 Maximum number of neurons

We thus need a method which suppresses these drawbacks in order to allow the implementation of a large number (hundreds) of neurons on a single chip.

THE PROPOSED CIRCUIT

Introduction

The main purpose of the proposed circuit is to suppress these drawbacks to allow the full interconnection of hundreds of neurons on a single chip. In order to eliminate the mismatching between the P-type and N-type current sources, all the synaptic currents will be sourced by N-type transistors only, but the positive currents will be sourced on one line and the negative currents on another one. Two different sums are then realized (one for the positive currents and one for the negative ones); the function of the neuron is to compare the total currents on the two different lines; the neuron will switch on if the total positive current is greater than the total negative one; otherwise it will switch off. The new values for each neuron are then fed back into the network through the different synapses. When all of the neuron values don't change anymore, a stable state is reached and the set of the neuron values forms the output of the network.

Synapses

Since a N neuron fully interconnected network requires N^2 synapses, the main part of the chip area will be occupied by the array of connections. In order to increase the neuron density on the chip, we will thus try to reduce as much as possible the synapse area. The logic diagram of a synapse is reproduced in Figure 4.

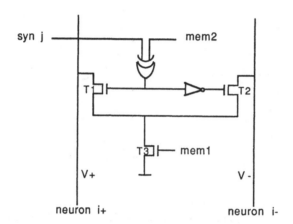

Figure 4 Synapse

As we saw previously, two values are stored in each synapse: "mem1" represents the absolute value of the connection while "mem2" represents its sign. The synapse can thus take the three different values -1, 0 and +1.

The output of the neuron j to which the synapse is connected is multiplied by a XOR function with the content of "mem2". This determines the sign of the current. "Mem1" commands the gate of a current source to determine if a current is to be sunk in this synapse. It is obvious that a single synaptic current has to be as small as possible because N currents can be summed on a neuron input. To avoid voltages in mem1 other than 0 and 5V, we use a long transistor (T3) as current source. The three permitted different combinations of mem1 and mem2 allow thus to sink a current on neuron i+, on neuron i-, or to sink no current at all.

Neurons

The function of the neuron is to compare the two differents currents on lines neuron i+ and neuron i-. First, the two currents are converted into voltages by transistors T4 and T5 (Figure 5). These voltages are themselves compared in the reflector formed by T6 to T9.

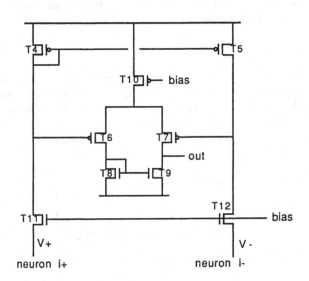

Figure 5 Neuron

Since a large number of synapses can be connected to the same neuron, we must add two transistors (T11 and T12) to keep the voltages V+ and V- as fixed as possible. Indeed, without these transistors, these voltages would decrease when we increase the number of active synapses; with a great number of neurons, we would quickly reach a point where the addition of one synapse will have no more effect on the neuron current because of the V+ and V- voltage diminution. Figure 6 shows the voltages on the output line (out) when there are N/2 active synapses on line neuron i+ and N/2+1 on line neuron i-, or vice versa. If we add a buffer to this output, we see that at least 500 synapses are allowed on the same line.

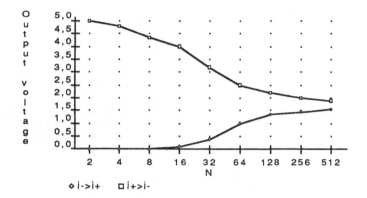

Figure 6 Neuron discrimination

Global architecture of the circuit

As described in the previous section, the proposed architecture can be used to implement a large number of neurons. However, to verify the theoretical predictions and to measure the computation speed of the network, we designed a 3x3 mm test chip in a CMOS 3 micron technology. The chip we describe here is a fully interconnected Hopfield network with only 14 neurons and 196 synapses.

Two memory points are contained in each synapses. To program this memory, a decoder selects the address line (i.e. the column) in which we want to write; the 28 memory points are then programmed simultaneously through 28 input pads. In the computation mode, 14 of these 28 pads are used as the outputs of the neurons. The loop between the neuron outputs and the synapse inputs can be interrupted to compute the neuron values step by step for test purpose.

Layout

Figure 7 shows the layout of the 14 neuron test circuit. It is divided in 14 neurons (1), 196 synapses (2), the RAM decoder (3), 14 input/output pads (4) used either to program the RAM or as output of the circuit and 14 input pads (5) used to program the RAM.

The chip has been realized in a CMOS 3 microns technology with single metal and single poly. Obviously the use of a large chip with a smaller technology will increase the number of neurons which can be put together on a single chip.

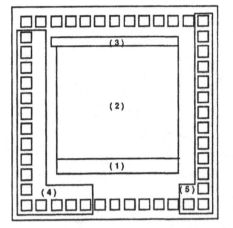 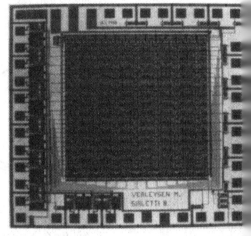

Figure 7   Layout

CONCLUSION

We have described a new VLSI architecture for the implementation of neural networks. Classical VLSI approaches present an important problem: computation errors appear when we increase the number of connected neurons. To suppress these drawbacks, we designed neurons and synapses where the comparison between currents is much more accurate than in circuits based on N-type and P-type current sources. This architecture allows the implementation of a fully interconnected network with hundreds of neurons, or a layered network with many more neurons.

We suppose here that only three values are allowed for each connection; the scope is to minimize the synapse area, since only two memory points are necessary to store the connection strength. However, classical CAM algorithms such as Hebb's rule need more values for each connection. There are two solutions to this problem: either we increase the number of possible strengths in each synapse, but its area will also increase, or we develop new algorithms more adapted to this architecture (it is possible to obtain better results than the Hebb's rule even with only three allowed connection values (Sirletti *et al.*, 1988)).

To design very large arrays of synapses, we are obliged to reduce their areas. We think that the use of different memory points, for example DRAM in place of SRAM, will considerably reduce the synapse area. We can compare the actual state-of-the-art in neural networks with the early developments of memories and look forward to very useful arrays of neurons with millions of synapses...

Acknowledgement

M. Verleysen and B. Sirletti acknowledge the support of IRSIA.

References

Fahlman, S. E. and Hinton, G. E., "Connectionnist Architectures for Artificial Intelligence", *IEEE Computer*, pp. 100-109, Jan 1987.

Graf, H.P. and de Vegvar, P., "A CMOS Implementation of a Neural Network Model", in *Proc. 1987 Stanford Conference in Advanced Research in VLSI*, pp. 351-367, 1987.

Hebb, D.O., *The Organization of Behavior*. New York: Wiley, 1949.

Hopfield, J.J., "Neural networks and physical systems with emergent collective computational abilities", in *Proc. Natl. Acad. Sci. USA*, vol. 79, pp. 2554-2558, April 1982.

Howard, R.E., Schwartz, D.B., Denker, J.S., Epworth, R.W., Graf, H.P., Hubbard, W.E., Jackel, L.D., Straughn, B.L. and Tennant D.M., "An Associative Memory Based on an Electronic Neural Network Architecture", *IEEE Transactions on Electron Devices*, vol. ED-34, pp. 1553-1556, 1987.

Lippmann, R.P., "An Introduction to Computing with Neural Nets", *IEEE ASSP Magazine*, pp. 4-22, April 1987.

Nicollian, E.H. and Brews J.R., *MOS: Physics and Technology*. New-York:John Wiley & Sons, Inc.,1982.

Personnaz, L., "Etude de Réseaux de Neurones Formels, Propriétés et Applications", Doctoral Thesis, University of Paris, June 1986.

Rumelhart, D.E., McClelland, J.L. and the PDP Research Group, *Parallel Distributed Processing*, vol. 1&2. Cambridge: MIT Press, 1986.

Sirletti, B., Verleysen, M. and Jespers, P.G.A., *A New Learning Algorithm for Content-Addressable Memories Using Hopfield's Neural Networks*, UCL Internal Report, April 1988.

Tsividis, Y. and Satyanarayana, S., "Analogue Circuits for Variable-Synapse Electronic Neural Networks", *Electronics Letters*, Vol. 23 N° 24, November 1987.

7.3 A LIMITED–INTERCONNECT, HIGHLY LAYERED SYNTHETIC NEURAL ARCHITECTURE

Lex Akers, Mark Walker, David Ferry and Robert Grondin

INTRODUCTION

Recent encouraging results have occurred in the application of neuromorphic, *ie.* neural network inspired, software simulations of speech synthesis, word recognition, and image processing. Hardware implementations of neuromorphic systems are required for real-time applications such as control and signal processing. Two disparate groups of workers are interested in VLSI hardware implementations of neural networks. The first is interested in electronic-based implementations of neural networks and use standard or custom VLSI chips for the design. The second group wants to build fault tolerant adaptive VLSI chips and are much less concerned with whether the design rigorously duplicates the neural models. In either case, the central issue in construction of a electronic neural network is that the design constraints of VLSI differ from those of biology (Walker and Akers 1988). In particular, the high fan-in/fan-outs of biology impose connectivity requirements such that the electronic implementation of a highly interconnected biological neural networks of just a few thousand neurons would require a level of connectivity which exceeds the current or even projected interconnection density of ULSI systems. Fortunately, highly-layered limited interconnected networks can be formed that are functionally equivalent to highly connected systems (Akers *et al.* 1988). Highly layered, limited-interconnected architectures are especially well suited for VLSI implementations. The objective of our work is to design highly layered, limited-interconnect synthetic neural architectures and develop training algorithms for systems made from these chips. These networks are specifically designed to scale to tens of thousands of processing elements on current production size dies. The network will be functionally equivalent to single or multi-layered fully interconnected architectures.

We report in this paper a limited-interconnect, highly layered perceptron-like network. The constraints and opportunities of VLSI technology drive the style of the architecture. A compact analog cell provides both the modulation of the inputs by the analog weights, and sums the products. A pull-down gate provides for shunting inhibition. The weights are stored dynamically on the gates of transistors thus allowing rapid pipelined processing. First, simulations of the limited-interconnect architecture are discussed demonstrating its ability to generalize and tolerate faults. Next, the analog cell and chip architecture for a 512-element, feedforward neural IC are described. Important timing and reset requirements are also discussed.

SYSTEM SIMULATION

The limited-interconnect, feedforward network was simulated in order to determine the performance limitations imposed by using a fixed topology architecture for different tasks. The network was trained in separate runs to perform three different operations on an 8-bit digital input vector using the back-propagation algorithm. The tasks implemented included inversion, right shift, and unsigned one's complement addition of a constant (binary 7). Highly structured operations of this type are not good candidates for neural network implementation (Abu-Mostata and Psaltis 1987). They are however, good "worst-case" algorithms for evaluating this architecture in that they require a one-to-one mapping (8-dimension input space to 8-dimension output), and the input vectors are highly correlated. In addition, since all possible responses are known, the accuracy of inferred outputs can be easily determined.

The total number of connections within an adaptive filter determines the degrees of freedom available to the net in forming hypersurfaces in decision space which implement the desired I/O mapping (Widrow and Stearns 1985). A design constraint for limited-interconnect, multi-layered networks therefore is that they must contain at least as many connections as the smallest, fully-connected structure which implements the same function. A fully-connected, two-layer perceptron (one layer of hidden units) is generally considered to be the smallest structure capable of generating arbitrary mappings (Plaut *et al* 1986). In these tests, a 2-layer net with ten hidden units was the smallest fully-connected structure which implemented all three tasks with an acceptable amount of error. In addition, because of the limited ability to generate dense interconnect patterns with double-metal CMOS design rules, a fan-in of four was selected as the maximum number of inputs possible at each cell without having to resort to exotic channel-routing schemes between adjacent layers. The final design of the limited-interconnect structure, which had approximately the same total number of connections as the fully-connected structure and maintained the required fan-in, was a 4-layer net with three hidden layers of ten units each. More precise design methods for feedforward networks which form distributed internal representations are not yet available. Only upper and lower bounds on network size have thus far been considered in the literature (Baum 1987, Hecht-Nielsen 1987, Lippman 1987, McClelland 1986). The actual minimum number of connections and processing elements required for a specific problem is a function of the relative degree of randomness in the desired mapping (Baum 1987).

The first set of results demonstrates the ability of the limited-interconnect structure to generalize untrained outputs from a limited set of examples. Figure 1 shows the response error as a function of training. Even though the three tasks for which the network was trained have very different governing algorithms, each network inferred untrained output in an almost identically linear fashion.

The second simulation demonstrates the relative fault tolerance of the limited-interconnect architecture. In this experiment, an increasing number of randomly selected connections within the network were selected and broken (weight set to zero). Figure 2 demonstrates that the functioning of the networks performing inversion and shifting was maintained in spite of a small number of broken connections. The actual amount of error introduced in the output however, depended on the particular connection broken and the learned task. Using the back-propagation algorithm, the relative importance of each connection in generating the desired mapping is randomly determined. In addition, since we are employing a fixed network topology for different tasks, the amount of redundancy within the net depends on the function implemented. For these reasons, many of the weights may contribute little to the desired behavior of the network and may be eliminated without significantly degrading performance. The network performing addition however, did not survive the loss of a few connections. The number of connections provided apparently did not allow for redundancy. Greater fault tolerance might have occurred with a larger network.

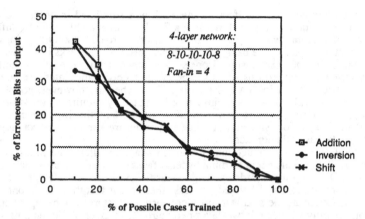

Figure 1 Network response error as a function of training

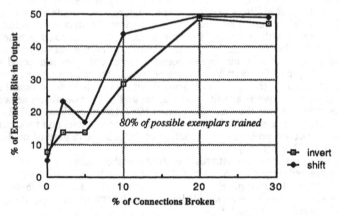

Figure 2 Network response error as a function of failed connections

LIMITED-INTERCONNECT CHIP DESIGN

The number of devices on a die has increased in a exponential fashion over the last 28 years. Regretfully, the number of interconnection layers has not increased at this rate, causing serious restrictions on information flow in VLSI and ULSI systems (Ferry *et al* 1988). The levels of interconnections have increased from only two to five. The increase in interconnection density, defined as the sum of the inverses of each metal layer pitch, plus the inverse of twice the polysilicon pitch (Myers *et al* 1986), has not increased sufficiently to support large, highly interconnected systems. This has important consequences for hardware implementations of synthetic neural networks. We must design cells and networks which do not require high interconnectivity when scaled to large numbers of processing elements. By designing a limited-interconnect architecture, one eliminates global interconnections by using local interconnections and multiple layers. This is an excellent trade off for VLSI technology and allows large systems to be rapidly designed and implemented by replicating the cells in both the row and column direction. Only the select logic needs to be modified on the die. This type of architecture also allows for system expansion with multi-chip sets.

Figure 3 is the circuit diagram of our limited-interconnect analog neural cell and Figure 4 is the timing diagram. The operation of the cell is as follows. Weights are stored dynamically on the gates of transistors T1, T2 and T3. Notice only PMOS transistors (T18, T19, and T20) are used to pass and isolate the weights instead of transmission gates. This is allowable since only weights above the device threshold are important, and hence a degraded low state voltage has no effect on circuit performance. For inputs of 5 volts, the drain parasitic capacitors of T7, T8, and T9 are charged by current flowing through the pass transistors T4, T5, and T6, to a voltage equal to the weights minus the device threshold voltage. For inputs of 0 volts, the pass transistors will allow the capacitors to discharge. While exact multiplication of the input and the weight is not done, shifting of the circuits' logical threshold voltage and modifying the training algorithm compensates for this behavior. In fact, one of the very useful characteristics of neural networks is exact generation of products are not necessary. Once the storage capacitor in each branch is charged, clock Φ1 is turned off to isolate the signal from the input. Turning on Φ2 allows the signals to be analog summed and compared to the logical threshold of the first inverter. The first inverter needs to be of minimum size to allow acceptable charge transfer. The circuit on top of the PMOS pull-up device allows the logical threshold and hence the neural threshold to be set at a voltage lower than .5Vdd. The output inverter restores the output voltage level and drives the next stage. Since this circuit uses only positive weights, a shunting transistor T11 is used to provide inhibition. We have proven that arrays made with this cell can perform as a complete logic family. Figure 5 shows a circuit simulation of the cell.

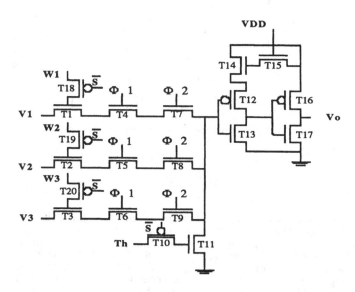

Figure 3 Analog synthetic neural cell

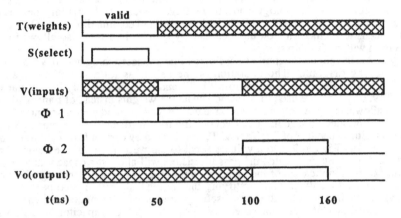

Figure 4 Timing Diagram for analog neural cell

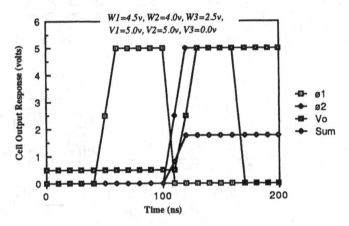

Figure 5   Circuit simulation of cell

A system formed by replicating the analog cell operates in the following manner. To keep the number of I/O pins at reasonable numbers as the system is expanded, we share the input, weights, and output pins. For the implementation of 512 neurons in an array 32 wide by 16 long, 32 lines are used for the weights, and I/O. The four unique weights are multiplexed to the single line running to each cell. Hence, a row is selected for a write, then weights for each cell are written. This continues for four cycles after which all weights for the row addressed have been written. The next row can then be selected for weight loading. The whole array can be loaded in approximately 10μs, an order of magnitude faster than the discharge time constant for an individual gate. Figure 6 shows the block diagram for the chip. For efficient data flow through the layers, the clock lines are interchanged in every other row as shown in Figure 7. After the weights have been loaded,

Address

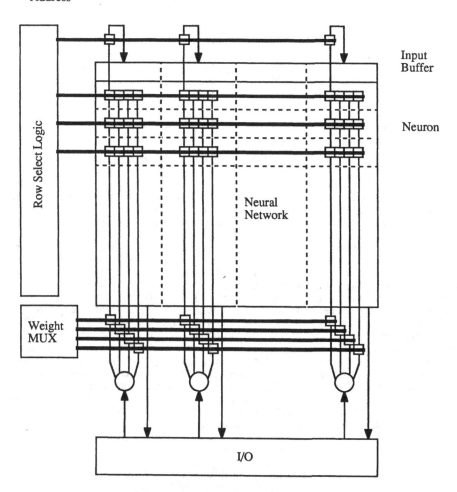

Figure 6 Block Diagram of Chip

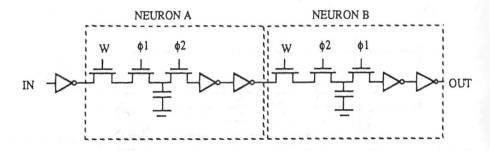

Figure 7 Two layers cells showing pipelining

the inputs are loaded, and the output vector ripples through the layers. A potential danger with storage capacitors is a weight value below the device threshold with a low input will not allow the capacitor to discharge. We have eliminated this problem by pipelining a wave of zeros through the network.

Figure 8 shows the layout of the synthetic neural cell with 3 micron p-well design rules. This layout style was chosen for ease of replication. All of the signal lines except the input and output signals are routed on vertical or horizontal metal lines through the cells. The connection matrix used in the network determines the number of rows necessary and the ability of the network to make complex decisions. Simulations have shown that the presence of edges in the network can cause instabilities. Signal propagation edges occur if the interconnections are not fully implemented on the physical edges of the circuit. There are two methods in forward propagation connection matrices to eliminate array edges. The first method is to run a connection from one side of the chip to the other. Clearly, as the array size is scaled up, this approach is impractical since the propagation delay would slow the chip down. The second method uses a connection scheme where the connections form a cylinder in three dimensions (Figure 9), and hence does not have an edge. This interconnect pattern can be mapped into two dimensions by renumbering the nodes. Figure 10 shows the resulting two dimensional interconnection scheme. Using this architecture, a 512-element, feedfordward neural IC has been designed.

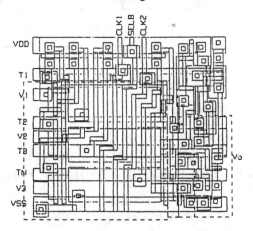

Figure 8 Layout of synthetic neural cell

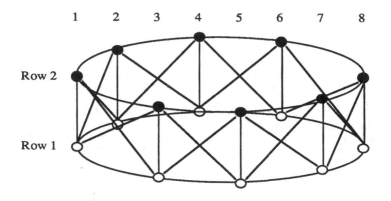

Figure 9 Interconnection Scheme in Three Dimensions

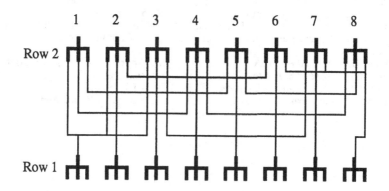

Figure 10 Two Dimensional Interconnection Scheme

CONCLUSION

We have designed and simulated a limited-interconnect synthetic neural IC. The network can simulate a complete logic family and performs like a fully connected network. A compact analog neural cell is developed which is compatible with a DRAM fabrication line and drives a style of architecture which is fully compatible for large scale integration.

ACKNOWLEDGEMENTS

The authors would like to thank Synergy for valuable contributions to the design, simulation, layout, and test of synthetic neural circuits. Synergy stands for **S**ynthetic **N**eural **E**ngineering **R**esearch **G**roup. This group includes W. Fu, C. Goh, P. Hasler, D. Hohman, W. Looi, A. Shimodaira, and T. Zirkle.

REFERENCES

Abu-Mostafa, Y.S., and Psaltis, D., "Optical Neural Computers," *Scientific American*, vol. 256, no. 3, March, 1987.

Akers, L.A., Walker, M.R., Ferry, D.K., and Grondin, R.O., "Limited Interconnectivity in Synthetic Neural Systems", in Rolf Eckmiller and Christopher v.d. Malsburg eds., *Neural Computers*. Springer-Verlag, 1988.

Baum, E.B. "On the Capabilities of Multilayer Perceptrons", *IEEE Conference on Neural Information Processing Systems - Natural and Synthetic*, Denver Colo., November, 1987.

Ferry, D.K., Akers, L.A., and Greeneich, E., *Ultra Large Scale Integrated Microelectronics*. Prentice-Hall, 1988.

Hecht-Nielsen, R., "Kolmogorov's Mapping Neural Network Existance Theorem," *Proceedings of the IEEE First International Conference on Neural Networks*, vol. 3, pp. 11-12, 1987.

Lippman, R.P., "An Introduction to Computing with Neural Nets," *IEEE ASSP Magazine*, p14-22, April, 1987.

McClelland, J.L., "Resource Requirements of Standard and Programmable Nets," in D.E. Rummelhard and J.L. McClelland eds., *Paralell Distributed Processing - Volume 1: Foundations*. MIT Press, 1986

Myers, G.J., Yu, A.Y. and House, D.L., "Microprocessor Technology Trends," *Proceedings of the IEEE*, Vol. 74, No. 12, p. 1605, Dec. 1986.

Plaut, D.C., Nowlan, S.J., and Hinton, G.E., "Experiments on Learning by Back Propagation," Carnegie-Mellon University, Dept. of Computer Science Technical Report, June, 1986

Walker, M.R., and Akers, L.A., "A Neuromorphic Approach to Adaptive Digital Circuitry," *Proceedings of the Seventh Annual International IEEE Phoenix Conference on Computers and Communications*, p. 19, March 16, 1988.

Widrow, B., and Stearns, S.D., *Adaptive Signal Processing*. Prentice-Hall, 1985.

7.4 VLSI–DESIGN OF ASSOCIATIVE NETWORKS

Ulrich Rückert and Karl Goser

INTRODUCTION

Associative Networks (ANs) and other similar information processing networks now experience an increased interest within different areas of computer science as well as within microelectronics. Many different models of such networks have been discussed in the branch of neuro–science (Anderson and Rosenfeld 1987). Software simulations have shown that these ANs store and process information effectively. The following attractive characteristics of these networks are based on the distribution of processing power amongst the data storage devices to minimize data movement:

- Associative recall of information means the reconstruction of stored patterns if the input only offers a portion or a noisy version of these patterns.
- Tolerance towards failures in the hardware means that losses of devices in the network cause only a slight decrease in the accuracy of the recall process, but do not affect the total function of the network.
- Parallel processing offers a concept at which every device in the network is doing something useful during every operation whereas in a conventional microcomputer a fast processor performs instructions very quickly, but the memory is idle during any instruction cycle.

The basic operations of ANs are pattern mapping (heteroassociative recall) and pattern completion (autoassociative recall). Associative networks can also solve optimization problems (Hopfield and Tank 1985). These subjects and further interesting features of such networks like fault–tolerance, generalization, and selforganization are discussed in the literature.

At present, the research of AN models is focused mainly on theoretical studies and computer simulations. However, if ANs should offer a viable alternative for storing and processing information in applications such as pattern recognition and classification, these systems will have to be implemented in hardware. The research on hardware concepts for ANs is just at the beginning. There are two different approaches for supporting ANs on parallel VLSI hardware, the design of digital accelerators and the design of special–purpose emulators. It is quite obvious that the relatively slow simulations of ANs on serial computers can be speeded up considerably by parallel hardware. The implementation by means of microprocessor controlled networks, for example array processor systems, is a promising compromise between flexible modelling –these networks are still program controlled– and a complete parallel processing of large matrices. The economy of time is proportional to the number of processors employed.

Perfect parallelism is achieved by special–purpose VLSI–systems, which are orders of magnitude faster than accelerators at the expense of flexibility. Due to the fast progress in microelectronics VLSI chips of these networks are feasible now.

This paper describes the basic design principles behind a silicon implementation of AN–emulators. These principles, which offer the basis for a wide variety of network types, are illustrated by a general VLSI architecture combining analog and digital CMOS techniques. In detail the function of ANs, the design towards a functional integration and the features of the VLSI architecture will be described. In this context an effective interaction between both system design and VLSI technology is important for realizing an AN as a microelectronic component successfully.

GENERAL STRUCTURE OF ASSOCIATIVE NETWORKS

An Associative Network is composed of many simple processing units (Figure 1). Each processing unit must be small in order to accommodate many on a chip, and communication must be local and hardwired. All processing units operate in parallel and are structurally equal, they have the same number of inputs. This assumption is important for a formal analysis as well as a technical implementation, not for the functionality itself. The external inputs x from the environment and the internal inputs y from other processing units as feedback form the total inputs e. The network output z is a subset of y.

The input e_i influences the output y_j of a processing unit by a connection weight w_{ij}, which should be programmable in a certain range. The connection weights can be excitatory ($w_{ij} > 0$) or inhibitory ($w_{ij} < 0$). A pair of equations shown in Figure 2 characterizes the dynamic behaviour of an associative network: The *transfer function* g of a processor unit and the *adaptation function* f (learning rule) of the connection weights. In general, the transfer function g of almost all AN types is based on the weighted sum of inputs of the processing units.

Different processing units and therefore types of associative networks result from the choice of the transfer function g. For example, Kohonen's Correlation Matrix Memory makes use of a linear transfer function, Hopfield Networks of a sigmoid function and Perceptrons of a threshold function. In general, the implemented function g is a strong simplification of the transfer function of a neuron. Though the biological transfer function is more complex and the details of it not yet fully understood, the weighted sum of inputs is widely accepted as a first order approximation.

In general the adaptation rule f is a function of the system parameters, too. In regard to a VLSI implementation only "local" adaptation rules will be considered in the following. In this case the change of a connection weight is proportional to the signals locally available at the connection element. The values of the connection weights w_{ij} as well as the output signals y_j can either be continuous or discrete.

The transfer and adaptation dynamics take place on different time–scales. A common assumption is that the rate of adaptation must be much slower than the dy-

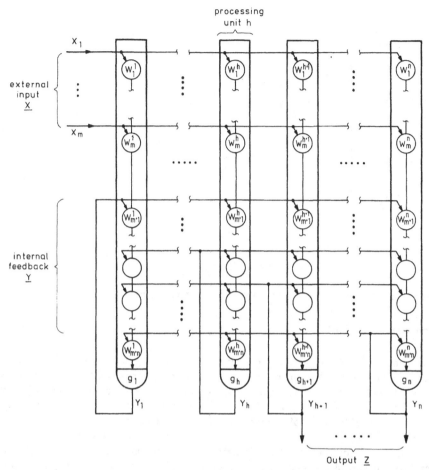

Figure 1 General structure of an Associative Network

namical response of steady network states, the so called stored or memorized patterns. Hence, upon a constant input the transfer function has to force the network quickly into a stable state. In this stable state the adaptation of connection weights takes place. As a consequence, many AN models distinguish between a learning phase, in which the adaptation to a given pattern set occurs, and a recall phase, in which the associative recall occurs.

This general framework contains almost all mathematically formulated associative networks. A concrete version of an associative network is obtained by a proper choice of the initial connection matrix and the dynamic functions for adaptation and propagation. However, this framework is very coarse and many degrees of freedom are left, but it reflects the highly parallel, regular and modular architecture of ANs making them attractive for VLSI system designer. An additional highlight is the ease of programming, done by adaptation and not by a program written outside.

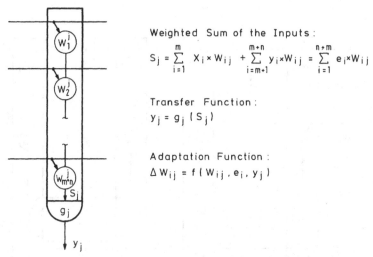

Weighted Sum of the Inputs:

$$S_j = \sum_{i=1}^{m} x_i \times W_{ij} + \sum_{i=m+1}^{m+n} y_i \times W_{ij} = \sum_{i=1}^{n+m} e_i \times W_{ij}$$

Transfer Function:
$$y_j = g_j (S_j)$$

Adaptation Function:
$$\Delta W_{ij} = f(W_{ij}, e_i, y_j)$$

Figure 2 General description of a processing unit

DESIGN TOWARDS FUNCTIONAL INTEGRATION

Analog versus digital

The main module of an AN is the processing unit consisting of individual connection weights for each input and an output circuit g. An appropriate circuit implementation of the function g is very important. In our opinion, the analog implementation is attractive because of size, power and speed. The pure digital AN emulator serially calculates the weighted sum of inputs and requires a data–bus per processing unit, having a width proportional to the data–format of the connection and input values. As a consequence, digital emulators must be synchronous and cannot utilize the inherent parallelism of associative networks. In the analog case the weighted sum of input signals can be computed by summing analog currents or charge packets for example. In Figure 3 a simple circuit concept is proposed in CMOS technology. The activation function S_j depends on the ratio of activated excitatory (k) and inhibitory (l) connections, which are both binary in this simple example. This ratio of the input signals has been transformed into a voltage. The transfer function g can then be performed by a conventional operational amplifier, e.g. the simplest circuit is a CMOS inverter or an analog comparator. Though this simple circuit is only a rough implementation of the weighted sum of inputs, it is already applicable to associative VLSI networks because of its compactness.

The accuracy of analog circuits is not as high as for digital circuits, but more appropriate for highly parallel signal transfer operations inherent in ANs. Therefore almost all IC's for this application up to now make use of analog computation of the transfer function.

The most critical task certainly is the integration of the connection elements. In literature several realizations have been proposed, ranging from non–programmable

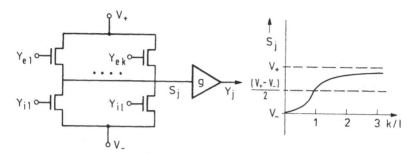

Figure 3 MOS–voltage–divider implementation of the activation function

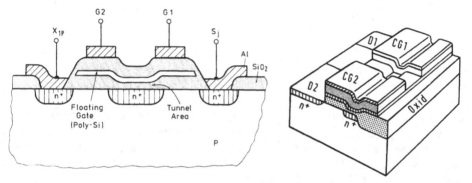

Figure 4 Schematic cross–sections of the connection element

binary to programmable analog interconnections. Whereas the design of binary connections is straightforward, the design of multi–valued connections has to balance
the cell size and the resolution of the connection weight. Furthermore, almost all
VLSI implementations are non–adaptive, but learning or self–organization requires
incremental adjustment of the weights. Therefore two concepts are proposed in the
following: a pure analog and a hybrid digital/analog concept.

The connection element as a floating gate transistor

As already proposed in earlier publications (Rückert and Goser 1987), the so–called
floating–gate transistor offers a good way for a functional integration of a connection element. Based on this transistor a new connection element with dual drain
and dual control gate has been designed as depicted in Figure 4. The device acts
as a non–volatile storage cell at which the electrical charge on the floating gate represents the information. Since the charge is only quantized in electrons the storage
is analog. It holds the stored information independently of the power supply of the
cell. The floating–gate transistor principle is mainly used in electrically erasable and
programmable read only memories (EEPROMs).

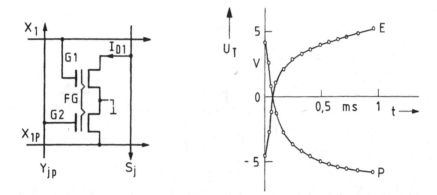

Figure 5 MOS connection element schematic and threshold voltage U_T
as function of the programming or erasing time

The cell matches the requirements for operating within a processing unit quite
well. During association an activated input line x_i causes the memory cell to influence
the column line sj by the current I_{Di} (Figure 5). The current I_{Di} depends on the
threshold voltage U_T, which is proportional to the stored charge on the floating–gate.
The current I_{D1} can therefore be adjusted by varying the charge on the floating gate.

When programming the transistor the threshold voltage is shifted by tunneling of
electrons into the floating–gate by activating either line y_{jp} or line x_{ip} with a programming pulse. The amplitude and duration of the programming pulse determine the
threshold shift and therefore the "adaptation dynamics". Using constant programming pulses e.g. with an amplitude of 15 V and a duration of 0.1 ms, the charge on
the floating–gate, representing the adaptive weight w_{ij}, can be changed according to
a characteristic shown in Figure 5 which resembles the adaptation function. At least
a resolution of 32 threshold shifts can be achieved according to first measurements.

The floating–gate transistor is not intended to store accurately an analog value,
because it is very difficult to program the cell precisely enough. The cell weight
is a measure of the importance of this connection between two processing units, or
in other terms the probability of this connection to be active. Though the storage
capacity cannot be enhanced by this implementation, the tolerance to faulty inputs
during learning can be achieved (Goser *et al* 1984) and self–organization as proposed
by Kohonen (1984) becomes possible.

In the present state of the art, the degradation effect due to charge trapping
within the thin oxide limits the satisfactory longtime operation of this device. After
around 10^7 modification cycles only a small threshold shift is left, which makes the
cell inoperable in a matrix. However, the trend of technology tends toward negligible
degradation by looking for better insulation layers.

The connection element with CCD loop

A connection element realized with charge coupled devices overcomes the disadvan-

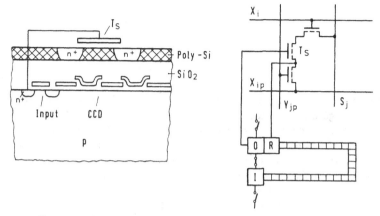

Figure 6 Cross–section and circuit schematic of the connection
element with CCD loop

tages of low accuracy and longtime degradation of the nonvolatile connection element
described above. The build up of the element is more complex since the element requires a three–dimensional integration with at least two layers of silicon as shown
in Figure 6. The 3D integration is necessary for saving area and having enough
possibilities for the wiring of the connection elements.

In the first layer, in the bulk substrate, a CCD loop is integrated together with a
regenerator and an input/output stage as well as the clock lines. The cycle time for
the CCD's depends on the leakage currents. The leakage rate can be reduced by three
orders of magnitude by cooling the circuits to –50 °C and by five orders of magnitude
by cooling to –100 °C. The number of CCD's in the loop yields the accuracy of the
connection weight since the number of charge packets corresponds to the weight value
and is only limited by the area.

The second layer is realized in a polysilicon film which is annealed by a laser
beam. In this layer the n and p channel MOS transistors represent the connection
element itself since the charge packets of the CCD loop are collected in their gate
capacitances: The higher the charge the lower the resistance or the higher the current
if the transistor is used in the current source mode. The power supply lines and the
input lines also have to be integrated on top of this layer.

The further advantage of this element is that the information stored into the loop
can be read out by opening the loops and by switching all loops in series. So the
chip output receives the charge packet strings of each connection element. From the
output this information can be stored in a memory, e.g. on a magnetic disc. On the
other hand the adaptive weights can be written in the loops from outside, so the time
consuming process of learning can be omitted if duplicating a system. However, this
concept is only superior to a digital version as long as the necessary resolution of the
connection element is limited. For current technologies this limit is 6 or 8 Bits.

The feasibility of this concept is high since the research work on 3D integration is
going on. In our laboratory we have developed a method for annealing a polysilicon
film in such a way that it is good enough for integration of the MOS transistors but

that the annealing temperature is low and the time short enough so the integrated circuits underneath are not affected by this process step.

STRUCTURAL DESIGN CONSIDERATIONS

For the VLSI system technique the highly regular and distributed structure of ANs results in a rigorous modularization of the complete system. Hence, as long as the whole system can be integrated on a single chip, the design of associative networks is straightforward as depicted in the above section. For practical applications, however, where the network has to be extended to a useful number of processing units, e.g. more than thousand, the whole network cannot be implemented on one silicon chip today. As a consequence, we have to look for an adequate partitioning of the network.

One of the decisive limiting factors for VLSI implementations of ANs is the pin requirement of each chip. The number of pins is mainly bound to the number of inputs and the number of processing units, respectively. Current VLSI packaging techniques limit the number of I/Os to a few hundred. In a straightforward way the network is vertically split into slices operating simultaneously, each of them controlling an equal number of processing units. This splitting strategy for large networks will not assure arbitrary extensions of the network because the number of inputs is fixed. A horizontal splitting of the network is problematic, because we have to transmit analog values among the chips.

For synchronously operating networks the problems diminish, since we can take advantage of multiplexing techniques and parallel–to–serial conversion via shift registers to conserve I/O–pins (Rückert and Goser 1988). An alternative solution for a special kind of binary AN with only a few activated inputs and outputs at any time have been presented by Rückert *et al* (1987). Assuming sparse input/output patterns it is appropriate to transfer the activated I/O–signals within the pattern by means of their addresses instead of the complete pattern.

For asynchronous ANs requiring direct connections between individual processing units these techniques are not applicable. Such systems distributed over several chips require one I/O–line for each input and output line of the processing units. As a consequence, silicon implementations of large asynchronous adaptive systems probably have to wait for the evolving Wafer Scale Integration. The entire network can be integrated on one silicon wafer, surrounded by conventional digital circuits taking over control and communication tasks. Because of their regular and uniform structure as well as their ruggedness against hardware defects ANs will be especially well adapted to this new integration technique.

The highly regular and modular structure of ANs is also indispensable for a successful management of the functional and testing complexity of future VLSI and WSI systems. Though sophisticated and special–purpose design tools are not available for hybrid architectures up to now, especially not for the higher functional levels, it is useful to build up a hybrid design data base for ANs. Such a design data base enables a non–specialized engineer to design a complete AN concept in only a few months, as demonstrated by several student projects in our laboratory. As a consequence, our research is concentrated on the design and fabrication of building blocks for ANs, based on a close interaction of system designer and technologists.

CONCLUSION

This paper describes design methods how to transfer associative networks into special–purpose VLSI–hardware. In our opinion, the point at issue is the implementation of an adaptive connection element, capable of learning from experience and by means of training. Up to now, there are only few proposals for connection elements in literature. Two of them are presented here, based on floating–gate MOS–transistors and on CCD loops. Both approaches have their advantages and it remains to be seen which type of circuit will be more effective in applications, and how closely a micro–electronic component of an associative network should resemble the highly interconnected nature of biological network.

Another important requirement is the compactness of the connection element, because the cell size mainly determines the overall area of the network. For a total matrix $n^2 + mn$ cells are necessary. Therefore a tradeoff exists between the integration level represented by the number of cells on a chip and the resolution of the connection elements and other circuitry. Hence, for large adaptive ANs an adequate partitioning is indispensable, which will especially influence the design of asynchronous ANs. Consequently, a functionally optimized implementation of ANs in silicon depends mainly on a close interaction of system design and technology. However, the impressive improvement achieved by putting an algorithm into silicon can only be done once. Further improvements will be closely tied to mainstream technological advances in such areas as device size reduction, new functional devices and wafer scale integration.

ACKNOWLEDGEMENT. The authors thank the DFG (Deutsche Forschungsgemeinschaft) for financial support.

References

Anderson, J.A. and Rosenfeld, E., *Neurocomputing: A Collection of Classic papers.* Cambridge Mass.: MIT Press, 1987.

Goser, K., Fölster, C. and Rückert, U., "Intelligent Memories in VLSI," *Information Science,* vol. 34, pp. 61–82, 1984.

Hopfield, J.J. and Tank, D.W., "Neural Computing of Decisions in Optimization Problems," *Biol. Cybern.,* vol. 52, pp. 141–152, 1985.

Kohonen, T., *Self Organization and Associative Memory.* New York: Springer Verlag, 1984.

Rückert, U. and Goser, K., "Adaptive Associative Systems for VLSI," in *WOPPLOT 86: Parallel Processing: Logic, Organization, and Technology,* J.D.Becker, and I.Eisele (ed), Berlin: Springer, pp. 166–184, 1987.

Rückert, U. and Goser, K., "VLSI Architectures for Associative Networks," in *Proc. of Int. Symp. on Circ. and Syst.,* pp. 755–758, Helsinki, 1988.

Rückert, U., Kreuzer, I. and Goser, K., "A VLSI Concept for an Adaptive Associative Matrix based on Neural Networks," in *Proc. of COMPEURO,* pp. 31–34, Hamburg 1987.

7.5 FULLY–PROGRAMMABLE ANALOGUE VLSI DEVICES FOR THE IMPLEMENTATION OF NEURAL NETWORKS

Alan Murray, Anthony Smith and Lionel Tarassenko

INTRODUCTION

A neural network is a massively parallel array of simple computational units (neurons) that models some of the functionality of the human nervous system and attempts to capture some of its computational strengths (see Grossberg 1968, Hopfield 1982, Lippmann 1987). The abilities that a synthetic neural net might aspire to mimic include the ability to consider many solutions simultaneously, the ability to work with corrupted or incomplete data without explicit error-correction, and a natural fault-tolerance. This latter attribute, which arises from the parallelism and distributed knowledge representation gives rise to graceful degradation as faults appear. This is attractive for VLSI.

Planar silicon technology is almost certainly *not* the ultimate medium in which neural networks will find their power fully realised. It is our view that to delay research into *implementation* of neural networks until analysis and simulation demonstrate their full power and a better technology emerges would be short-sighted. There is much to learn from LSI/VLSI implementation, and any hardware networks developed will be able to make rapid use of developments in network design and learning procedures to solve real problems.

NEURAL NETWORK ARCHITECTURE AND COMPUTATIONAL STYLE

This section discusses the architecture, signalling strategy, and computational style used, without reference to detailed MOS circuitry.

Overall Architecture

All 2-dimensional implementations of neural networks share a common general architecture. Neurons signal their states $\{V_i\}$ upward into a matrix of synaptic operators. The state signals are connected to an n-bit horizontal bus running through this synaptic array, with a connection to one synaptic operator in every column. Each column consists, therefore, of n operators, each adding a new contribution T_{ij} to the running total of activity for the neuron i at the foot of the column. The function of the neuron is therefore to apply a sigmoidal function to this activity

x_i to determine a neural state V_i. The synaptic function is to *multiply* a neural state V_j by a synaptic weight T_{ij} (stored in memory local to the synaptic operator), and *add* the result to a running total.

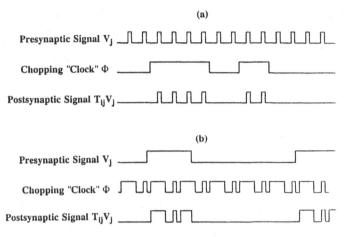

Figure 1 Chopping Clock Technique

This architecture has many attractions for implementation in 2-dimensional silicon :-

- The large summation $\sum_{j=0}^{n-1} T_{ij} V_j$ is distributed in space.
- The interconnect requirement (n inputs to each neuron) is distributed through a column, reducing the need for long-range wiring to an n-bit state "bus".
- The architecture is modular, and can be expanded or cascaded with ease.
- The architecture is regular.

Signalling Mechanism

We have given the name "pulse stream" to the signalling mechanism used by our neural circuitry. The process is analogous to that found in natural neural systems, where a neuron j that is "on" fires a regular train of voltage spikes (at a rate R_j^{max} pulses/sec) on its output (or axon), while an "off" neuron does not. We use exactly this signalling mechanism, in that one of our synthetic neuron circuits receives a weighted summation from its input synapses and operates upon this activity to decide a state, and a firing rate.

Arithmetic operates directly on these streams of pulses, with synaptic weights in the range $-1 \le T_{ij} \le 1$. The state of a neuron V_j is represented by a firing rate R_j, such that $R_j = 0$ for $V_j = 0$, and $R_j = R_j^{max}$ for $V_j = 1$. We may multiply the state by (say) one half (from $V_j = 1$ to $V_j = 0.5$) by removing half of the presynaptic pulses. Similarly, we can multiply by 0.25 by removing three quarters of the pulses and so on. The product $T_{ij} V_j$ therefore becomes the original pulse stream representing V_j, *gated* by a signal that allows the appropriate fraction of pulses through.

Figure 1 shows this with a neural state V_j. A "chopping" signal Φ is introduced that is *asynchronous to all neural firing*, and is logically "high" for exactly the correct fraction of time to allow the appropriate fraction T_{ij} of the presynaptic pulses through. In Figure 1(a), the chopping clock has a frequency well *below* $R_j{}^{max}$ and appropriately-sized bursts of complete neural pulses are allowed through. In Figure 1(b), each neural pulse is chopped by a signal that is of *higher* frequency than $R_j{}^{max}$.

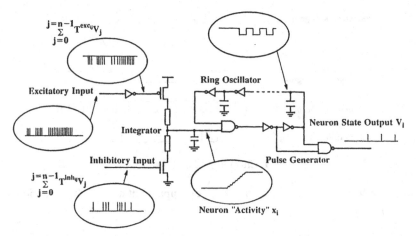

Figure 2 Circuit Diagram of Pulse Stream Neuron

Neuron Function

The neuron receives excitatory and inhibitory inputs, and produces a state output. If the neuron is initially "off", with relatively weak inhibition, the onset of stronger excitation turns the neuron "on", and it commences firing at its maximum rate R^{max}. It may be subsequently switched "off" by strong inhibition.

Synaptic Weighting Function

The synaptic function is also straightforward. The (positive or negative) synaptic weight T_{ij} is stored in digital memory. To form the product $T_{ij}V_j$, the pre-synaptic neural state is gated according to the chopping signals derived from T_{ij}. The resultant product, $T_{ij}V_j$, is added to the running total propagating down either the excitatory or inhibitory activity channel, to add one term to the running total, as shown. One binary bit (the MSBit) of the stored T_{ij} determines whether the contribution is excitatory or inhibitory.

NEURON AND SYNAPSE CIRCUIT ELEMENTS

In this section, the function blocks outlined above for neural and synaptic functions are expanded into MOS circuitry.

Neuron Circuit

Figure 2 shows a pulse stream neuron i. The output stage consists of a ring oscillator whose natural frequency is R^{max}, driving a "pulse generator", to convert the oscillator square wave to a sequence of short pulses.

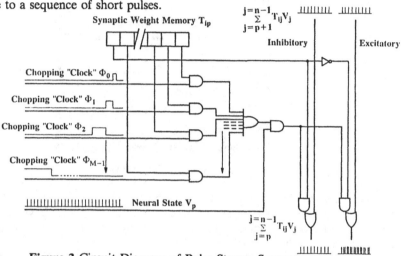

Figure 3 Circuit Diagram of Pulse Stream Synapse

The oscillator loop is broken by a NAND gate. The NAND gate acts as an inverter, completing the ring, if the neuron "activity", x_i, is 0V, and causes the oscillator to fire as x_i rises to 5V. The neural activity is represented by the voltage level on the capacitor on the NAND gate input.

To determine this activity level, the streams of aggregated inhibitory and excitatory pulses are applied to an "integrator" circuit. A p-channel transistor dumps a small packet of charge on the integrating capacitor whenever an excitatory pulse reaches its gate, while an n-channel device removes packets of charge when inhibitory pulses arrive at its gate. In the diagram, the excitatory pulses are more frequent (ie the excitation exceeds the inhibition), and the neural activity rises as more charge is dumped onto the capacitor than is removed from it. As a result, the neuron switches "on", and begins to fire.

Synaptic Weighting Circuit

Figure 3 shows a pulse stream synapse T_{ip}, with a precision of M bits. The M chopping signals "Φ_0, Φ_1,...Φ_{M-1}" are introduced to match the binary bits 0 to M-1 of the synaptic weight, while the "M^{th}" bit determines the sign of the weight. Clock Φ_{M-1} is high for 50% of the time, clock Φ_{M-2} for 25%, clock Φ_{M-3} for 12.5% , etc. The NAND gates attached to the weight bits will therefore allow 50%, 25%, 12.5% of the presynaptic pulses in V_p through, *if the corresponding bits of T_{ip} are logically high*. The chopping signals are asynchronous to the neuron firing signals, and the network dynamics, but synchronised to one another.

Figure 4 Chip Photograph

The chopping clock signals selected by the bits of T_{ip} are then OR'ed to form the total chopping clock, which gates the presynaptic neural signal V_p via an AND gate. The resultant product signal, $T_{ip}V_p$ is subsequently OR'ed on to the appropriate output channel, according to the MSBit of T_{ip}.

The chopping signals can be either much slower or much faster than the neural firing rate. Provided the aggregated pulse streams are integrated over a time constant much longer than either the chopping clock period or the firing rate period, it is the proportion of time during which the total input signal is high that matters. This will be the same in both cases, regardless of whether bursts of entire pulses or fragmented pulses are incident on the neuron inputs.

RESULTS

Physical Layout

Figure 4 shows a chip photograph, representing a section of the synaptic array. At present, the *neural* function is realised in discrete SSI (neuron) and custom VLSI (synaptic array) parts to allow maximum flexibility in choice of capacitor values, and therefore time constants. The chip integrates 64 synapses, each occupying 200μm x 400μm, so the total chip area is 16mm². It should be noted that the restriction on chip complexity in this application is fundamentally one of pin count, rather than of area. As Figure 4 shows, some silicon area is wasted, because the standard frame used precludes the availability of extra pins to support further synapses. Pin count can be reduced by multiplexing data into and out of the chip.

Simulation Results

Figure 5 shows a device level (SPICE) simulation of the neural circuit in Figure 2. (V4) is the integrator output, representing x_i. A strong excitatory input causes the neuron to turn "on", during which time the neural potential can be seen to rise in steps (corresponding to packets of charge being dumped on the integrator capacitor) until the ring oscillator begins to fire. Subsequently, a stronger inhibitory input removes charge packets from the capacitor at a higher rate, driving the neural potential down and switching off the ring oscillator. The "firing" pulses therefore cease.

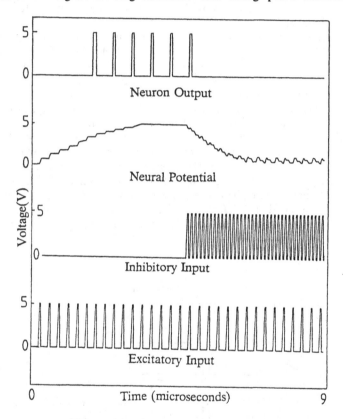

Figure 5 SPICE Simulation of Neuron

PULSE WIDTH MODULATION USING ANALOGUE WEIGHTS

We present an alternative signalling technique involving the modulation of pulse widths using analogue weights. This technique involves using a voltage controlled resistor (VRO). This resistor is used to control the discharge rate of an inverter which in turn modulates the width of an output pulse.

Figure 6 illustrates a synapse in such a system. The synapse has two elements. These are an inverter (M1-M2) with programmable discharge (pull-down) resistance (M3) and circuitry to allow analogue voltages to be stored on an internal capacitor.

Transistors M1, M2 and M3 constitute the voltage controlled inverter. When a pulse arrives at the input to M1 and M2, node Y is discharged. M3 acts as a voltage-controlled resistor. Since the voltage across C1 can be modified, the discharge rate can be modulated. By passing this sawtooth waveform through another inverter a second pulse can be recovered. The width of this pulse is determined by the point at which the waveform at Y goes below the switching threshold of the second inverter.

The analogue voltage used to control the discharge rate of the inverter is stored on capacitor C1, which is implemented in CMOS technology. Standard CMOS has the disadvantage that the capacitor is implemented largely by storing charge on the transistor gate. To overcome capacitance leakage without using large capacitors (which require large silicon areas) or a special fabrication process it is proposed that a refresh system similar to that used in DRAM's be used (although in this case it is an analogue voltage rather than a digital one which is being refreshed). This has the advantage that weights can be refreshed and even changed "on the fly" whilst the system is in operation. The weight values are stored in external RAM and are converted to the analogue voltage by a DAC before being "fed" to the chip. The precision of the weights can be changed by altering the width of the memory and the DAC. The internal capacitor is addressed via a transmission gate indicated as M4 in Figure 6. It is proposed that the chip will have its own internal refresh addressing system, with only the clock, reset and analogue weighting signal being provided to the chip in order to keep the pin count to a minimum.

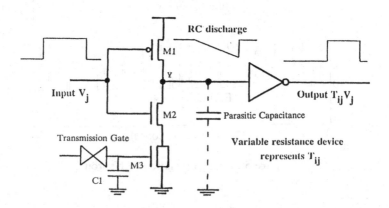

Figure 6 Circuit Diagram of Pulse Stream Weights

There are several departures from ideal behaviour. The non-linear doping across the chip surface leads to different resistive values for the M3 transistors for the same gate voltage. This can be overcome using the refresh system. Since weight values are stored externally, an offset value could be added or subtracted to compensate for this effect. The compensation values could be calculated from initialisation tests. Secondly there is the problem of mixing analogue and digital circuitry on the same chip. Digital circuitry can cause current spikes on the power supply lines whilst switching.

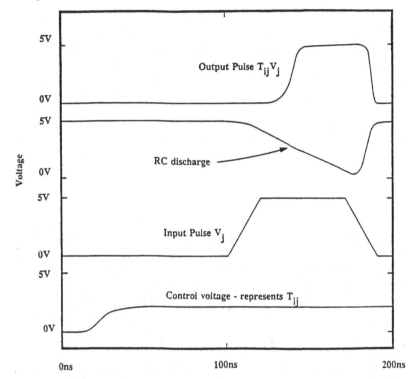

Figure 7 SPICE Simulation of Synapse with Analogue Weights.

Figure 7 shows the ouput from preliminary SPICE simulations from this circuit without careful device sizing. Initial simulations suggest that the control voltage will be in the region 1.7V to 2.7V. Below 1.7V the multiplication becomes non-linear and above 2.7V the transistor becomes saturated. This voltage range may change as the system is developed.

This synapse can be used as a direct replacement for the chopping synapse previously discussed. Further circuitry will be included to direct the pulse to either an excitatory or inhibitory line and memory will be included on chip to indicate whether the synapse is inhibitory or excitatory. The pulses will be OR'ed together and used to calculate a new neuron output as in the chopping system.

CONCLUSIONS AND FUTURE WORK

At present, a neural board has been assembled and interfaced to a host computer for loading weights and initiating computations. The board will comprise a small number of neurons initially ($\simeq16$) to test the technique properly, and to acquire some experience in controlling the dynamics of this unusual circuit form. Subsequent to this trial period, we will assemble a more significant pulse stream network computer. Initial results show that the pulse stream network can be used as a content addressable memory, and some progress has been made in using the Wallace learning algorithm (see Wallace 1985) for updating the weight set. Research is continuing into improving the neuron oscillators to minimise the number of discrete external components needed to control the oscillators. We are presently laying out the analogue weight chip, which it is hoped will be fabricated within the next twelve months.

The initial application area envisaged for our hardware is in automation of the Grossberg/Carpenter classifier network (see Carpenter and Grossberg 1987) although the "learning" portion of the network's behaviour will still be timestepped.

REFERENCES

Carpenter, G. A., Grossberg, S., "A Massively Parallel Architecture for a Self-Organising Neural Pattern Recognition Machine", in *Computer Vision, Graphics and Image Processing*, vol. 37, pp. 54-115, 1987.

Grossberg, S., "Some Physiological and Biochemical Consequences of Psychological Postulates", in *Proc. Natl. Acad. Sci. USA*, vol. 60, pp. 758 - 765, 1968.

Hopfield, J. J., "Neural Networks and Physical Systems with Emergent Collective Computational Abilities", in *Proc. Natl. Acad. Sci. USA*, vol. 79, pp. 2554 - 2558, April, 1982.

Lippmann, R. P., "An Introduction to Computing with Neural Nets", in *IEEE ASSAP Magazine*, pp. 4 - 22, April, 1987.

Wallace, D. J., "Memory and Learning in a Class of Neural Network Models", in *Proc. Workshop on Lattice Gauge Theory : A Challenge in Large Scale Computing*, November, 1985.

Chapter 8

ARCHITECTURES FOR NEURAL COMPUTING

Speed up of neural computing can be accomplished by means of dedicated computers. Although these computers do not have an architecture that resembles a neural network model, they do provide high computational power for neural computing applications. Properties of neural network models are implemented on these architectures by means of software/hardware algorithms. Architectures for neural computing are based on hardware that has been extensively studied; hardware such as random access memories, systolic arrays and microcomputer networks. Advantages of this approach may include flexibility, reduced connectivity, and functionality. In this chapter three novel architectures are presented.

THE RANDOM ACCESS MEMORY APPROACH

A neuron can be modelled as a system with a variable logic function between input and output. Such function can be expressed in terms of linear activation formulae that may be described by a variable truth–table. Aleksander §8.1 has developed a *probabilistic logic node* that implements the attributes of neural networks. This approach is implemented on conventional random access memories (RAMs). Truth–tables are stored on the memories; the weights of the inputs are also considered on these tables. Learning on this RAM based machine is described in the paper.

SYSTOLIC ARRAYS

Systolic array architectures present several attractive features that make them suitable for VLSI implementations. These features include: short intercommunications which allow minimum delay and high bandwidth, regularity that gives high packing density, and parallelism which may leads to high performance (Moore *el al* 1987). Blayo and Hurat §8.2 present a N^2 recurrent systolic array that implements N neurons. Convergence, threshold and input/output problems are considered in the paper. A 16×16 cell array has been designed; the cell architecture is described in detail.

MIMD APPROACH

Multi-instruction Multi–data stream (MIMD) machines can speed up computation of problems that can be efficiently partitioned and in which the communication overhead is not too large. In order to provide flexibility and task–independence, Garth and Pike §8.3 consider an MIMD architecture that could simulate neural networks. The system, GRIFFIN, has the following input parameters: the size and shape of the nets, the organization of nets into a system, the shape of the sigmoid, and the learning rate. The basis of the machine is a distributed array of autonomous neural network simulators.

References

Moore, W., McCabe, A. and Urquhart, R. (Eds.), *Systolic Arrays*. Bristol, England: Adam Hilger, 1987.

8.1 ARE SPECIAL CHIPS NECESSARY FOR NEURAL COMPUTING?

Igor Aleksander

INTRODUCTION

The McCulloch and Pitts model still reigns supreme in most schemes of neural computing and influences VLSI makers in their designs. This creates difficulties largely at the level of having to find analogue memories or their emulations, to satisfy the weight designs. Here it is shown that a logical analysis of the neuron leads to a logical model which avoids analogue memory and is centred on the conventional Random Access Memory. Defining the FUNCTIONALITY of a neural node as the number of distinct functions that the node can perform in response to training, it is argued that digital attempts at designing weights should be examined closely in comparison with direct logical methods to assess whether sufficient functionality is obtained for the silicon area used. Whatever the case, VLSI makers ought to be aware of the logic RAM alternative, as it is closer to the design of conventional digital chips for computing systems.

THE RAM-NEURON ANALOGY: FUNCTIONALITY.

Take the binary version of the McCulloch and Pitts model:

Output = 1 iff sum (over j from 1 to N) of w(j)*i(j) >T
   where there are N inputs to the neuron, i (j) is the value of
   the jth input (= 0 or 1) and w(j) is the weight of the jth input and T is some variable threshold.

For any given set of weights and T is is quite possible to draw the truth table for the element. For example, let w(1) = 1, w(2)=2, w(3)=3 and T=3. Then the truth table is:

Table 1 Truth Table

i(1)	i(2)	i(3)	OUTPUT
0	0	0	0
0	0	1	0
0	1	0	0
0	1	1	1
1	0	0	0
1	0	1	1
1	1	0	0
1	1	1	1

If one were to interpret the first three columns of the truth table as the address inputs of a random Access Memory (RAM) with a word length of only one bit, it is easily seen that the above function could be emulated simply by storing the right bits at the right addresses.

The functionality of the neuron model was a matter of interest in the mid 1960's and has been discussed by Muroga (1965). The neuron model can only perform linearly separable functions (i.e. those that may be achieved by an N-1 hyperplane of the N hypercube). Typical of the functions that such a device cannot perform are the parity checking ones that feature significantly in discussions on error back-propagation (Rumelhart, Hinton and Williams, 1986) On the other hand, the RAM can perform all the possible functions which number $2^{\wedge}2^{\wedge} N$ ($^\wedge$ reads 'to the power of'). The functionality of the two models for the first few values of N is shown below:

Table 2   Functionality

N	McCP	RAM
1	4	4
2	14	16
3	104	256
4	1882	65,536
5	94,572	4.3×10^9
6	15×10^6	18.5×10^{18}

The danger is that one might implement an M-input McCP model with W-bit weights and achieve a lower functionality that the $2^{\wedge}$ (M*W) offered by the RAM method.

NODE LEARNING AND GENERALISATION

A fear that arises when considering the RAM way of doing things is that comfortable learning schemes such as the well-trodden Widrow-Hoff rule need to be abandoned. Also it is precisely the restricted functionality of the McCP model that gives it a power of generalisation which, for the RAM appears to have been lost. It will be shown below that such schemes retain both these properties.

To take an example, consider the system that gives rise to TABLE 1, above. Assume that it starts with its weight at zero, the threshold at 3 and that the object of the training task is to make the node fire whenever the majority of inputs (i.e. 2 in this case) is at 1. Much of what happens depends on the order in which the example input patterns are presented during training. But the problem could be seen as one of showing the node two prototype patterns 000 and 111, and expecting it to generate appropriate responses to those patterns nearest to the prototypes in Hamming distance (i.e. the difference in number of bits). But, remembering that a rule such as the Widrow-Hoff operates by distributing the responsibility for cancelling an error among the weights and the threshold, it may be calculated that the presentation of 111 would accomplish something like $w(1) = w(2) = w(3) =1$ and $T=2$. The result is that only 111 responds with a 1. Presenting now 110, say, the result may be something like $W(1) = W(2) = 1.3$ and $T = 1.7$. At this point the appropriate response is obtained for 101 and 011 which exceed the T as well at 100, 010, 001 which do not. This is what is meant by 'generalisation'.

The scheme suggested by Aleksander (1988) shows that exactly the same expectations may be fulfilled if a low amount of noise is used during the training the RAM node. Again 000 and 111 may be used as the prototypes of the task, and generalisation obtained by training not only on these prototypes, but also versions slightly distorted by noise. Indeed, it has been shown that for RAMs, it is possible to define an ideal RAM content with respect to the training instances, which guarantees best generalisation. But more of this below, where a slight modification to the RAM model is proposed.

THE PROBABILISTIC LOGIC NODE (PLN)

Although the McCP model is still salient in much of neural computing, recent work has shown that a modification of this model is necessary if hard learning tasks are to be achieved. One such mode is that used with feed-forward nets and training by error back-propagation (Rumelhart *et al* 1986).

The essence of such models is that they provide an analogue output computed according to some SQUASHING FUNCTION of the inputs. That is if

S= sum over j of w(j), i(j) (i (j) now being a real number)

then, OUTPUT = $f (S) where $f is the squashing function which is any differentiable function in S.

Usually, $f is an S-shaped function with origins in the Fermi-Boltzmann formulation that was used in Hinton's Boltzmann machine (Hinton and Sejnowski, 1986). The output number is seen as lying in the interval (0,1) and in the Boltzmann machine, this was the probability of firing a 1 in response to S.

The difference between these squashing function formulations and the binary McCP model allow the weight changing regime to be such as to ensure first, that in dynamic nets (i.e. nets with feedback), the training creates low energy states through simple local measurements across the weights even for the hidden units that are necessary in hard learning tasks. In feed-forward nets although the output is now merely a real number, the reason for the squashing function again relates to the control of hidden units.

Put simply, the squashing function first allows the node to be set to a 'don't know' state for some or all the inputs (OUTPUT = 0.5, say). It also allows the training algorithm to place rank-order on the inputs in terms of being associated with outputs in the 0-1 continuum.

The PROBABILISTIC LOGIC NODE (PLN) is endowed directly with these simple properties. In contrast with the single-bit RAM model, the input now addresses a b-bit word and outputs a b-bit number in the interval 0-1. We call this number B. Since the inputs to the node are still binary, the closure between the output of one node and the input to another, can be acieved by turning B into a probability of firing at the output of the node. We call this P(1). In practice, this can fire if the random number is less than B as shown in Figure 1.

A simple LOCAL TRAINING RULE becomes obvious: rewards consists in incrementing the value of B towards 1 if it is greater than .5 and towards 0 if is less than .5. Punishment means incrementing always towards the .5 value. Various graded methods can be used to do this. Precisely as in squashing function models this allows the outputs to be associated with 'don't know' states and to be rank-ordered against probability of causing firing. But there is an added property. Due to the lack of generalisation of the node, the rank-ordering is done independently for the individual inputs. This gives a faster convergence.

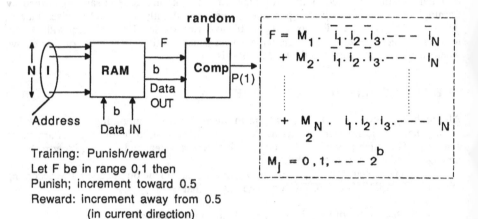

Training: Punish/reward
Let F be in range 0,1 then
Punish; increment toward 0.5
Reward: increment away from 0.5
(in current direction)

FIGURE 1   The PROBABILISTIC logic node (1987)

In a particular version of this arrangement, reported in Aleksander (1988), B is only given the values 0, 0.5 and 1. This gave a remarkably fast convergence to the 'parity problem' as defined in Rumelhart, Hinton and Williams (1986). But to discuss this, we must first look at the structural issues and the training of hidden units in particular.

PYRAMIDAL STRUCTURES

In Figure 2 we see the way in which the PLNs might be formed into a pyramid. Such pyramids have the property of generalisations and indeed, may be trained by an overall reward/punishment scheme. All the nodes in the pyramid are rewarded whenever the apex outputs the correct value and punished whenever it fails to do so consistently. Several detailed algorithm are possible under such a regime, they differ only in the definition of what constitutes proper and erroneous output. An example of a simple scheme that demonstrates generalisation is given below.

Take a three-layer pyramid with N=2 and D=4 hence W=8, using the symbols of Figure 2. Also assume the 3-level version of B. Assume that all the B values start off at 0.5.
Let the task be one of detecting whether the left or the right of a string of 8 bits contains an adjoint group of 1s among 0s. Purely intuitively we let the two training patterns be

      1 1 1 1 0 0 0 0 to fire with P(1) of the output node at 1 and,
      0 0 0 0 1 1 1 1 to fire with P(1) of the output node at 0.

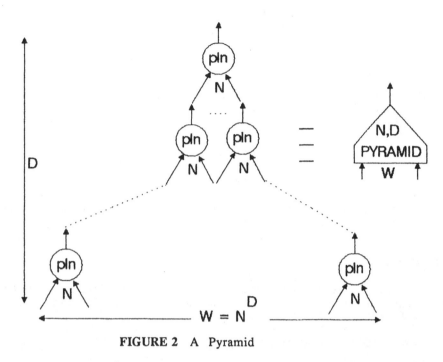

FIGURE 2 A Pyramid

We can represent the firing probabilities at the three levels by showing the input level first and the others below. For example, the response of the net to the first training pattern, before any training takes place may be shown as follows:

$$
\begin{array}{ccccccccc}
1 & 1 & 1 & 1 & 0 & 0 & 0 & 0 \\
& .5 & & .5 & & .5 & & .5 \\
& & .5 & & & & .5 & \\
& & & .5 & & & &
\end{array}
$$

This simply says that before training, all parts of the pyramid have equal probability of firing. We apply the training rule that as soon as the pyramid provides the correct output, it is 'jelled': that is, all the current values of the element outputs are firmly associated with their inputs (because of the three-level nature of B). So without loss of generality we can assume that the jelled values for the first training patterns are:

$$
\begin{array}{ccccccccc}
1 & 1 & 1 & 1 & 0 & 0 & 0 & 0 \\
& 1 & & 1 & & 1 & & 1 \\
& & 1 & & & & 1 & \\
& & & 1 & & & &
\end{array}
$$

The second training pattern will have a set of jelled internal representations that is largely arbitrary with respect to the first. It may be something like:

0 0 0 0 1 1 1 1
1   0   0   1
1   0
0

Now, given a previously unseen test patten as shown below, it is possible to calculate the firing rates of the nodes on the basis that a previously unseen input will cause the node to fire with P(1) = .5

1 1 1 0 0 0 0 0
.5   1   1   1
.75   1
.875

This shows generalisation at least to the extent that the pyramid is sensitive to the relative difference of the unseen pattern to the trained ones.

PRABABILISTIC RULE

A simple rule may be used to predict the output of a node P(out) given that the probability of its inputs firming in a trained way are p(1), p(2).......pN, respectively. It may be shown that

$$\text{P(out)} = 0.5 * [1 + p(1)*p(2)*.....*p(N)] \tag{1}$$

ADDITIONAL GENERALISATION

The addition of the probabilistic element only introduces the generalisations that the RAM lacks by providing the continuum in which similarity may be detected. However, two other factors need to be taken into account: first that the constrained functionality of a pyramid as opposed to a single RAM implies generalisation. Second, as has been pointed out in the case of a single PLN, using a small amount of noise during training, sets up functions in the RAMs that are close to an ideal (in terms of maximum generalisation for that particular training set). Both of these effects may be illustrated with a pair of examples. Consider two nets:

A:N=2,   D=4, W=16
B:N=4   D=2, W=16

We first assume that there is only one training pattern: the pattern of 16 1's for which the output is forced to 1. We call this experiment X. A second experiment (Y) has the same training pattern plus an additional training of all 16 patterns that contain only one 0. This represents training with a noise level of 1/16. The results for experiment X are listed in Table 3 below.

Table 3 Pyramid Behaviour in Experiment X

Hamming Dist	Net A response %	Net B response %
0	100	100
1	94	75
2	88	63
3	84	56
4	81	53

So far it is possible to note that the value of N is a powerful factor in determining generalisation and its complement, discrimination. The higher the N the greater the discrimination: a characteristic result for all logic versions of neural nodes.

The results of experiment Y are a little harder to predict as, it is necessary to bear in mind the distribution of input bits that differ from a training pattern. Also, there are several ways in which a particular training pattern could be jelled with respect to previous training patterns. However, it may be shown that a reasonable approximation in the case of net A, is that the response to a Hamming Distance 2 patterns will be 100% if both 0s occur in separate A-level nodes.

Table 4 : Noisy Training Results

Hamming Dist.	% Max. Response (%Confid.)	% Mn.Response (%Confid.)
	Net A	
1	100(100)	
2	100(94)	94(6)
3	100(80)	94(20)
4	100(67)	97(0.07)
	Net B	
1	100(100)	
2	100(57)	75(33)
3	75(100)	
4	75(60)	63(40)

In quoting the results we therefore quote a confidence with which a particular response is likely to be generated. Maxima and minima are reported to take into account different dispositions of the 0s.

Comparing this with the result in Table 3, it is indeed clear that noisy training leads to generalisation. Again, the net with a higher connectivity has greater discrimination. The entire question of generalisation may not be all that important when single pyramids are considered but, if say IxI pyramids are arranged so as to transform an IxI image into another IxI image and the system is then made dynamic by feeding the output image back to the input, it is this generalisation that creates attractors in state space.

CONCLUSIONS

Central to the discussion in this paper is that notion that care should be taken in the use of VLSI techniques to generate devices that aid the implementation of a neural computer. The reason for building such a machine must ultimatley be the achievement of some aspect of computing performance that cannot be expected for conventional computers. The key question is whether it is always wise to start with silicon models of neurons in the style of the McCulloch and Pitts suggestion.

The point that this paper is making is that the Probabilisitc Logic Node does not stray from conventional digital techniques and implements directly just those attributes that are found in a classical neural model: the ability to learn with controlled functionality and generalisation. Indeed through the introduction of a pyramid, it has been argued that a choice of pyramid characteristics allows the machine designer greater freedom in selecting generalisation and functionality than with a McCulloch and Pitts derivation.

The methodology used in this paper is by no means rigorous: that is thought to be beyond the scope of the aims of the paper. Only hints have been given of how the predictions are done. But others have eloquently provided comparative analyses. For example, Sherrington and Wong (1988) have shown that random nets made of PLNs have advantages over Hopfield models.

Clearly, it is not being argued here that the development of chips based on McCulloch and Pitts formulations should be abandoned, merely that the VLSI designer should bear the PLN alternative in mind, as it may be a more appropriate route towards the satisfaction of his aims.

REFERENCES

Aleksander, I., 'The Logic of Connectionist Systems' in Neural Computing Architectures, I. Aleksander (ed), North Oxford Press, 1988.

Hinton, G.E. and Sejnowski, T., 'Learning and Relearning in Boltzmann Machines', in Parallel Distributed Processing, vol. 1, D. Rumelhart and J. McClelland (eds), MIT Press, pp282-317, 1986.

Muroga, S., 'Lower Bounds for the Number of Threshold Functions and a Maximum Weight', IEEE Trans. on Electronic Computers, April 1965

Rumelhart, D., Hinton, G.E. and Williams, R.J., 'Learning Internal Representations by Error Propagation', in Parallel Distributed Processing, vol. 1, D. Rumelhart and J. McClelland (eds), MIT Press, pp 318-362, 1986.

Wong, M., and Sherrington, D., 'Storage Properties of Boolean Neural Networks', Proc. nEuro 88, Paris, 1988.

8.2 A VLSI SYSTOLIC ARRAY DEDICATED TO HOPFIELD NEURAL NETWORK

François Blayo and Philippe Hurat

INTRODUCTION

The VLSI technology has increased the complexity and performances of computers. Even extending this technology to its extreme of Wafer Scale Integration, problems still appear beyond our computational abilities such as real-time vision, hand-writing and speech recognition. Massively parallel architecture (Kung 1982) is one way to solve these complex problems.

On the other hand, Artificial Intelligence aims at solving these problems. AI algorithms aim to build internal representations of an environment. Up to now, the classical approach of AI has been a formal manipulation of symbolic expressions computed on von Neuman architectures. But, internal representations of an environment can be built by a massively parallel structure : neural networks.

As simulations of neural networks are time consuming on a sequential computer, dedicated architectures are well suited to implement these networks. Even commercially available hardware accelerators are painfully slow for large neural network simulations. The best solution consists in designing dedicated VLSI chips (Graf *et al* 1988).

Among the several proposed neural models, the Hopfield model has rather good performances in pattern recognition. Moreover, it is a well-established and simple model. Thus the integration of such a network is advisable. Hopfield neural models work in two steps: learning phase and recognition phase.

First, a learning procedure produces the synaptic weight matrix which is a synthesis of prototype patterns. Different learning rules have been developed (Hopfield 1982, Personnaz 1985). For a specific application, a learning rule produces an appropriate synaptic matrix.

After this initial step, the synaptic matrix is repetitively used to recognize input patterns: i.e. to match them with the prototype patterns. This recognition procedure is performed many times. A programmable recognition VLSI network would allow it to be run with good performance.

This paper presents such a device, dedicated to pattern recognition based upon the Hopfield neural model. The main drawback of this single-layer network is that neurons must be fully interconnected. The solution we present consists of implementing an N neuron network as a systolic square array made up of N^2 cells.

To perform the recognition procedure, a recurrence equation must be computed. A single step of the recurrency can be easily implemented on a data flow systolic array. We have defined a suitable data propagation performing the full recurrent recognition algorithm. Moreover convergence must be detected. Located on the west side of the array, a systolic device detects a vector convergence. When a vector has converged, a dedicated device proceeds to the suitable data I/O at the east side. The array efficiency is maximum when 2N vectors are simultaneously recognized. The complete architecture has been developed and integrated using ES2 CMOS 2μm technology.

RECOGNITION PRINCIPLES

The learning procedure produces the synaptic matrix which we call C. This matrix can then be set into a programmable recognition network and recognition can then be performed many times with good performance. An unknown pattern to be matched is shown to the network at time zero by forcing the network input. After this initialization step, the network iterates in discrete time steps, using the following formula :

$$X_{i,r} = F(\sum_{j=1}^{N} C_{i,j} \cdot X_{j,r-1}) \tag{1}$$

where:
$X_{i,r}$ is the ith component of the input vector X at the iteration step r,
$C_{i,j}$ is the synaptic matrix,
F is the threshold function,
N is the number of neurons.
The network is considered to have converged when outputs no longer change from iteration to iteration i.e:

$$\forall\, i \in [1,N],\, X_{i,r} = X_{i,r-1}. \tag{2}$$

The pattern specified by the network outputs after convergence is the recognized pattern.
Hopfield (1982) has proven that this network used with binary inputs converges when :
- the weight matrix is symetric ($C_{i,j}=C_{j,i}$)
- the threshold function is the output sign.
If we assume that the synaptic weights are stored in a NxN cell array, a suitable data propagation can perform the recognition algorithm.

SYSTOLIC ARRAY DESCRIPTION

Principle

We started from a classical systolic algorithm performing the matrix-vector product: the synaptic matrix components are set into the systolic array cells while vector components propagate step by step through the array. A single recurrence step of the recognition algorithm can be easily implemented on such an array (Blayo and Hurat 1987, 1988).

Each cell (i,j) is a small computational element which contains the synaptic weight between neuron i and neuron j: $C_{i,j}$. The X components propagate vertically from cell to cell. Each cell (i,j) performs the product $C_{i,j}.X_j$ and adds it to the subtotal received from the west cell. The complete data flow is shown in Figure 1.

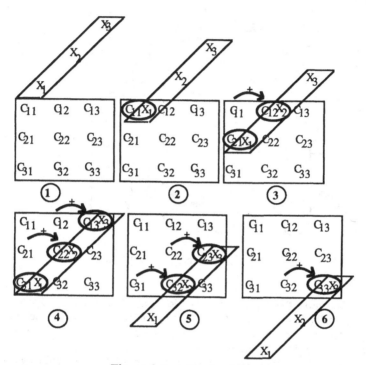

Figure 1 Systolic algorithm

Each row (i) computes $\sum C_{i,j}X_j$. A 2N step cycle is required to compute a single recurrence step where N is the length of vector X. This systolic algorithm provides $\sum C_{i,j}X_j$ on the east side of the array. According to (1), these results must then be thresholded to become the new X components and the computation goes on until convergence is obtained.

Due to long interconnections between the east and the north side of the array, a direct implementation is infeasible. We describe a suitable data propagation to solve this problem.

Recurrent Systolic array

The principle of our data propagation is based on three main rules :
- A scheduling rule: computation scheduling is the same as shown in Figure1.
- A systolic rule: data reach cells when they are required for processing.
- A propagation rule: at the end of a recurrence step, the new X components replace the

old ones.

In order to conform to the last rule, the east side results are fed back into the array after being thresholded as shown in Figure 2.

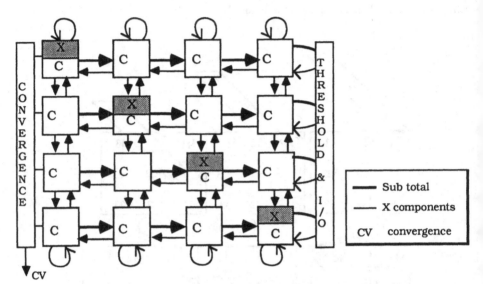

Figure 2 Recurrent data propagation

After initialization, the X components are located in diagonal cells. These components propagate vertically from cell to cell, first to the north and they are fed back into the array when they reach the north or south side. After 2N steps, the old X components reach the diagonal cells from which they started.

On the other hand, the subtotals are propagating from west to east until they reach the east side of the array. To become the new X components, they are thresholded and fed back into the array. After 2N steps, the new X components reach diagonal cells where the vector X components are updated to continue the recurrence.

CONVERGENCE AND I/O DEVICES

The recognition procedure must be performed until convergence is reached. The network output after convergence is the recognized vector. To detect the recognition convergence, the comparison $X_r = X_{r-1}$ must be done. A systolic device integrated at the array edges detects the vector convergence and proceeds to the appropriate data I/O.

In the diagonal cell $C(i,i)$ the comparison $X_{i,r} = X_{i,r-1}$ can be achieved because both the old and the new components are in the same diagonal cells. The comparison result is propagated to the west. The global convergence is computed at the west side of the network with a systolic AND function which produces a signal CV. This signal means that the first component of the recognized vector reaches the east side of the array. This component must be sent to the host computer and the first component of the new vector must be sent into the array.

The I/O device integrated at the east edge of the array achieves two different functions:
- If the component belongs to a non-recognized vector (CV=0) then the thresholded component is fed back into the array.
- If the component belongs to a recognized vector (CV=1), the ouput of the component and the input of a new vector component are achieved.

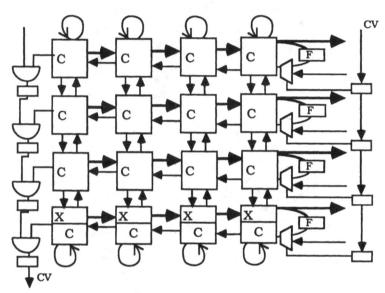

Figure 3 Convergence and I/O devices

We have described a recurrent systolic array performing both the recurent recognition algorithm and convergence detection. We must now examine the implementation efficiency.

PIPELINING

Our recurrent systolic array performs N^2 significant computations with N^2 cells in $2N$ time units. The activity ratio is then very low $1/(2N)$ because only a subset of cells performs a significant computation if the array only recognizes one vector at a time.

But we can improve the array activity without increasing the computational element complexity. The architecture allows the pipeling of $2N$ vectors into the array. All cells perform then a significant computation and the activity ratio reaches its maximum value 1. The convergence detection is done as before.

CELL ARCHITECTURE

The cell architecture of the previously described SIMD systolic array is given in Figure 4.

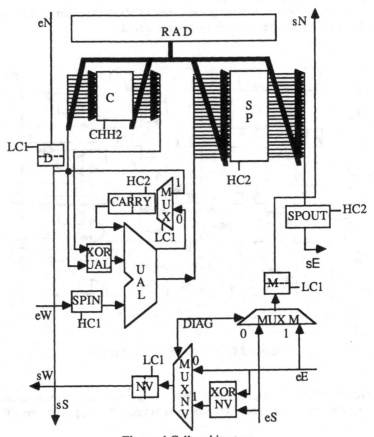

Figure 4 Cell architecture

Before computing recognition, synaptic weights must be loaded into the array. During the loading, the matrix components are presented to the array on its east side. A path can easily be managed from east cells to register D of each cell as shown in Figure 5.

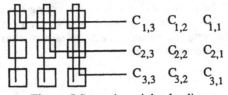

$$C_{1,3} \quad C_{1,2} \quad C_{1,1}$$
$$C_{2,3} \quad C_{2,2} \quad C_{2,1}$$
$$C_{3,3} \quad C_{3,2} \quad C_{3,1}$$

Figure 5 Synaptic weights loading

The signal CHH2 loads into register C the value stored in register D as shown in Figure 4.

During the recognition, each cell performs the same computation. A cell (i,j) receives from the North the Xj component, performs the product C(i,j).Xj, adds it to the subtotal received from the West and sends it to the East. Each cell must store components which are going both upward and downward, the subtotal, the fed-back thresholded value and the synaptic weight. Thus, each cell (i,j) must include 5 registers:

- $C_{i,j}$ which is the synaptic weight between neuron i and neuron j.
- SP which is the Subtotal.
- M which contains the value of $X_{i,r}$ which goes upward.
- D which contains the value of $X_{i,r}$ which goes downward.
- NV which allows the new value of $X_{i,r+1}$ to be fed back into the array.

Each cell performs the same computation i.e:

$$SP_{i,j} := SP_{i,j-1} + D_{i-1,j}*C_{i,j} \text{ with } SP_{i,0} = 0 \text{ and } D_{0,j} = M_{1,j}. \tag{3}$$

Hopfield networks are normally used with binary inputs (-1,+1). D contains the sign of an X component. Thus only an adder-substractor is required. The cell computation is then:

$$SP_{i,j} := SP_{i,j-1} \pm C_{i,j}. \tag{4}$$

But I/O bandwidth problem is a very critical problem in systolic arrays. To solve it, the transmission of SP values from cell to cell is performed on a serial mode. Thus computation is serial and a single bit adder-substractor can be used. The operation is defined by the D value which controls the two's complement adder through MUXUAL and XORUAL as shown in Figure 4.

A control part is required to handle the serialisation of the computation in a cell. Different devices have been studied. A serialisation control achieved by a shift register (RAD) has been chosen because it is the most flexible : size of register SP is programmable. RAD points the position of the SP and C bits under processing. This position is 1-out-of-n encoded in the shift register RAD.

Two different clocks are used to control the cell :

- a High frequency Clock signal (HC) which is the shift signal of the control part. It controls the SP values transmissions i.e SP, registers SPin and SPout.
- a Low frequency Clock signal (LC) which is the reset signal of the control part and of the two's complement adder. It defines the beginning of the cell computation. It also controls the communication part i.e. NV, D and M value transmission.

Clock frequencies are defined according to the number of significant bits in register SP (SIZE (SP)) by the following rule :

$$freq (HC) = (SIZE (SP) + 1) * freq (LC) \tag{5}$$

SIZE(SP) is defined according to the application. According to (4), SIZE(SP) depends on the number of significant bits of register C (SIZE(C)) and the number of neurons (N) by the following rule :

$$SIZE (C) + Log_2(N) \leq SIZE (SP) \tag{6}$$

The register SP sends a bit to the adjacent cell where it will be processed. But we must keep in mind that the cell will be integrated on chip and that the global array will be made up of several chips. The connection between two adjacent cells may be done through chip I/O pads. The register SP must then be able to load both SP output bus and I/O pads. To reduce the transmission delay, the reading of SP value is anticipated in the SPout register : the $i+1^{th}$ bit of SP is stored in SPout while the ith bit is computed.

As shown before, diagonal cells have a special purpose. They are different as far as communications are concerned and they perform the comparison $X_{i,r,k} = X_{i,r+1,k}$. Two multiplexors MUXNV and MUXM driven by a DIAG signal carry out these differences.

At the chip level, the local DIAG signals of diagonal cells are connected to the chip DIAG signal but the local DIAG signals of non-diagonal cells are pulled down to ground.

At the network level, the DIAG signals of all chips along the network diagonal are pulled up as shown in Figure 6. This flexible programming allows to build large networks by assembling several identical chips.

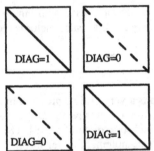

Figure 6 DIAG signal programming

The complete architecture shown in Figure 4 and described just before has been integrated in a VLSI technology.

INTEGRATION

First we designed one cell; the size of synaptic weights depends on the number of neurons. Our estimation shows that for N neurons $Log_2(N)+1$ bits are required. As we aim to build a 128 neurons network, the register C must then be 8 bit long : $SIZE(C) \leq 8$. The format of register SP is then : $MAX(SIZE(SP)) = 8 + Log_2(N) = 15$. The register SP has been implemented as a 16 bit register.

This architecture has been implemented, the floor plan and layout are shown in Figure 7. The circuit was integrated using the 2μm double metal CMOS technology of ES2 with the french MPC. The size of the operating part of a cell is 230μm x 490μm i.e. 0,1 mm^2 for 376 transistors. We designed a cell with a shift register. The circuit contains 600 transistors in 0,17 mm^2. The design density is higher than 3500 Tr/mm^2. The complete cell has been electrically simulated. The clock frequency estimation is 30MHz.

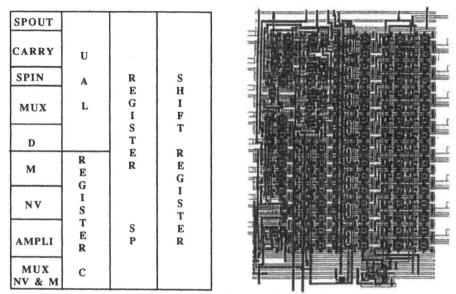

Figure 7 Floor plan and layout of the integrated cell

We are now proceeding to the test of the integrated cell. If the test confirms simulation results a 16x16 cell chip will be integrated in the same technology. A 128 neuron board will be built by assembling 64 chips before eventually studying a Wafer Scale Integration. With such a device about 200 million neuron updatings can be performed each second.

CONCLUSION

The data propagation described here allows the implementation of a large single layer network based on the Hopfield recognition algorithm. For a specific application, an appropriate synaptic matrix can be set into our programmable network and recognition can be performed many times with good performance. Due to the regularity of the defined SIMD array cell, large networks can be built. We are now integrating a 16x16 cell array on chip. With such a chip, we aim at building a 128 neuron board performing about 200 million neuron updatings each second. Moreover, large and powerful neural networks might be built as a wafer system as the integration of such a 2D systolic network is well-studied (Ivey *et al* 1987).

REFERENCES

Blayo, F. and Hurat, Ph., "Approche systolique pour une architecture neuro-mimétique", Technical report No29, IMAG, LCS, Grenoble, 1987.

Blayo, F. and Hurat, Ph., "Intégration d'une architecture systolique neuro-mimétique", in *Proc.* journée LAMI, E.P.F.L., Lausanne, 1988.

Graf, H.P., Jackel, L.D. and Hubbard, W.E., "VLSI Implementation of a Neural Network Model", IEEE Computer, Vol 21, No3, pp. 41-49,1988.

Hopfield, J.J., "Neural networks and physical systems with emergent collective computational abilities", in *Proc. Natl. Acad. Sci.*, USA, Vol 79, pp.2554-2558, 1982.

Ivey, P., Huch, M., Midwinter, T., Hurat, Ph. and Glesner, M., "Design of a large SIMD array in Wafer Scale Technology", *IFIP Workshop on W.S.I.*, Brunel Univ., 1987.

Kung, H.T., "Why systolic architecture ?", *IEEE Comp.*, Vol 15, No1, pp 37-46, 1985.

Personnaz, L., Guyon, I., Ronnet, J.C. and Dreyfus, G., "Character recognition, neural network and statistical physics", in *Proc. Cognitiva 85*, Paris, 1985.

8.3 AN INTEGRATED SYSTEM FOR NEURAL NETWORK SIMULATIONS

Simon Garth and Danny Pike

INTRODUCTION

One of the major problems with neural networks research is the immense quantity of computation required on digital computers to simulate analog networks. A number of workers have chosen to address this problem by implementing the structures in integrated circuit form and have gained very large degrees of speed-up in this way (Jackal et al 1986). However, such strategies are generally highly directed towards a particular task and a given mode of implementing that task. As a means of demonstrating real applications of a technique or permitting work to be done on a different aspect of a problem, the method has no rival. However, it is less well suited as a basis of a more general purpose vehicle for the investigation of artificial neural systems.

Apart from their lack of speed, digital computers have many advantages. The ability to reorganize the characteristics of the simulated system at will and the exceptionally good visibility which they afford into the inner workings of the problem make this approach most attractive as a method of investigating such systems. Regrettably, the lack of speed cannot be ignored, as run times of CPU days or weeks on conventional minicomputers are common for even moderately sized problems.

A natural compromise is to apply parallel processing to the problem. The inherently parallel nature of neural networks problems makes the task considerably simpler than that of adapting an arbitrary (necessarily serial) program from a digital computer to such a machine.

The purpose of the current development has been to build a machine which can provide the equivalent computational power of a supercomputer in the specific case of simulating neural network systems, at the approximate cost of a minicomputer. This is achieved by taking advantage of the inherent regularity in neural network simulations and reflecting this in the design of the hardware.

THE MODEL

In order to efficiently match the machine to the problem, it is necessary to define the model of the problem with some care. The systems under consideration consist of units (*neurons*) whose inputs are connected to the outputs of other units. The excitation of each unit is determined by forming the product of its inputs and the weights (*synapses*) between those inputs and the unit. This sum of products is then operated on by a non-linear function (typically a *sigmoid*) to produce the output of the unit, which then connects to a number of other units via *synapses* and the process repeats. Systems may be recursive or non-recursive as desired.

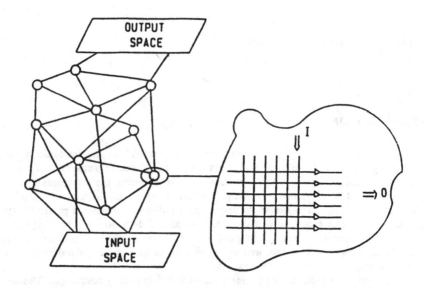

Figure 1 The organisation of nets in the GRIFFIN

One of the major difficulties in such arrangements is the large number of connections required between units. Rather than forming these connections explicitly, which would be both complex and inflexible, the processing units are conventionally organized into a regular structure (e.g. a 2-dimensional matrix or a hypercube) and messages are passed between units to simulate the interconnection (Hillis 1986). However, even with this scheme, if a separate processor is used for each unit, the system will rapidly become communication bound, despite the use of higher order communication structures.

A simplification is to regularize the structure into fully interconnected nets. A net containing n inputs and n outputs contains n*m weights but needs only to make n+m external connections. In addition, nets may be easily dedicated to a given sub-function which, although not necessarily optimally efficient, can considerably improve the ability of the researcher to attach *meaning* to the results of the learning process.

The basis of the system under consideration in the GRIFFIN consists of a number of fully interconnected nets whose inputs and outputs may be joined to arbitrary places within the system or to the external inputs/outputs (figure 1). The logical interconnection map is fully defined in software and may be easily modified to simulate different structures.

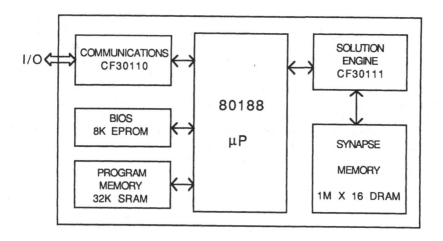

Figure 2 A block diagram of the NETSIM card

In all such simulators, there is a trade-off between efficiency and flexibility. However, in most cases, two functions may be identified as those which ultimately limit the speed of neural network simulations: the formation of the sum-of-products of outputs and weights and the updating of weights during the learning phase. The former may be represented as a repeat (multiply-sum) operation. The latter is commonly chosen to be a variation on the delta rule which corresponds to a repeat (multiply-sum-update) operation. These operations have to be performed once for each of the weights in the net, whereas most of the other operations (such as computation of the non-linear function) occur only once for each unit. Therefore, under most circumstances, the two operations described above dominate the computation time. There are a number of variations on these computations (such as the introduction of momentum in the synaptic update stage) but most of them may be implemented as a sequence of the above operations.

Flexibility needs to be provided in a number of regions. In particular, the size and shape of the nets, the organisation of nets into a system, the shape of the non-linear function and the learning rate are all key parameters in any such simulation. Thus, the requirement is for a general-purpose processor with a small number of specialist functions to expedite the most time consuming parts of the simulation while retaining the flexibility and familiarity of a more conventional processor. This is the rationale behind the design of the GRIFFIN system.

THE NETSIM

The basis of the machine is a distributed array of autonomous neural network simulators (figure 2). Each NETSIM consists of a local microprocessor, a solution integrated circuit (a specialist co-processor to implement the neural network functions at high speed) and a communications integrated circuit to allow large numbers (tens to hundreds) of NETSIMs to be connected together to form the parallel processing GRIFFIN (Garth 1987).

The solution integrated circuit is an autonomous vector processor attached to up to 2 MBytes of memory which holds the variables for the neural network simulation (weights, input vectors, etc.). The timing of the memory is linked to the vector processor, so that conventional low-cost DRAM circuits may be used despite computation rates (i.e. a vector dot product operation) of approximately two million *synapses* per second in the current machine (depending on the size and shape of the net in question). Each instruction includes the data-fetch and loop operations as well as the multiply-and-sum and replaces several instructions on a more conventional processor. Similarly, a multiply-sum-update operation is provided, including a prescale to adjust the learning rate, with a maximum computation speed which is a third of the multiply-sum rate. For the purposes of simulation, weights are stored as analog terms represented by 16-bit integers and the inputs and outputs as 8-bit integers.

Most of the inner-loop operations can be performed by these functions, or a combination of them. However, the microprocessor also has access to the DRAM contents so that (on principle) any operation may be undertaken explicitly by the local microprocessor, albeit at lower speed.

The vector processor is independent of the local microprocessor and appears to its as a register file in its address space. Having initiated the vector operation, the microprocessor is then free to compute other elements of the simulation, including the non-linear function and the simulated interconnection. The resultant information is then loaded into the communications integrated circuit for transmission to the relevant node in the system.In this way, a very high degree of flexibility is maintained (most parameters of interest are defined in software), while incorporating the speed improvement of a specialist processor.

The communications integrated circuit is a 3-dimensional serial message-passing device. The coarseness of the system (i.e. the degree of processing at each node) is such that even a powerful simulator will contain a relatively small number of nodes (the prototype machine has 36 nodes, equivalent to some 36 million *synapses*). This has a number of consequences. Firstly, there is little benefit to be gained from very high order interconnection systems between the nodes (e.g. a binary hypercube) since, with smaller numbers of nodes, the difference between high and low order systems in terms of the maximum distance a message must travel to reach its destination is relatively small (e.g. a factor of 3 for a 1000 node system between a binary hypercube and a conventional 3-dimensional cube). As a result, a 3-dimensional system has been chosen, as this has natural parallels with the layered planes style of architecture which is common in neural network solutions, particularly to visual problems. Any logical interconnection scheme may be achieved by adjusting parameters in software. Secondly, the coarseness helps to apportion the computation task amongst the three main processing elements (the microprocessor, the vector processor and the communications chip) so that no one element dominates the solution time.

The NETSIM is a complete neural network simulator with memory, local microprocessor, high-speed vector processor and a communications device. As such, it may be used in its own right as a simulator for a single neural network (or a number of smaller networks partitioned within the local memory) or may form part of a larger simulation machine.

THE GRIFFIN

The GRIFFIN is a parallel processor consisting of a distributed array of NETSIM simulators (figure 3). In order to improve the useability of the system, it is desirable to base such processors on existing architectures for which there is a large software base. The NETSIM card consists of an Intel 80188 and a quantity of RAM and EPROM which may be used for programming the local microprocessor. Included in the EPROM is an IBM-compatible BIOS which provides default returns for inappropriate calls to it (e.g. it returns *keyboard buffer empty* to a keyboard access). This allows a certain amount of commercially available software which runs on IBM-compatible computers to be run in the remote nodes, notably high level languages, such as Borland International's *Turbo C.* This considerably simplifies program development.

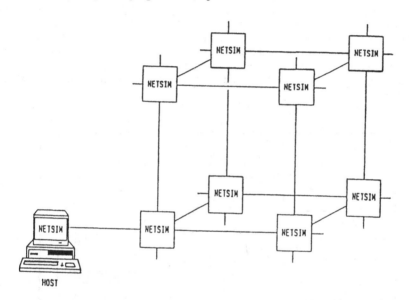

Figure 3 The physical organisation of the GRIFFIN

A further corollary of this choice is that a standard PC may be used as a development environment for the NETSIM. This is achieved by creating a version of the NETSIM card in which the microprocessor and memory are removed and replaced by access to the PC's bus. Software may then be designed, developed and debugged on the PC using all the conventional programming and debugging tools. In addition, calls may be made to the PC's I/O devices during the simulation process for debugging purposes and the code may be subsequent downloaded and run in the remote nodes as the local BIOS will manage all illegal device calls. Finally, to provide further visibility into the machine, any NETSIM card within the cube may be removed and replaced with a PC-based NETSIM emulator. This will precisely emulate the function of the NETSIM and may be used simultaneously to analyse details of the progress of a simulation using the PC's tools in the context of other NETSIMs performing other parts of the simulation.

Communication may take two forms. The high speed serial communications link provides a means of communicating with a specific node using messages which contain data and a relative address tag. This is a two-way network and is used for messages between the nodes and from the nodes to the host. In addition, the host has another, quite separate, broadcast channel available. For this, each node is given an absolute address, which is its *key* (this may be loaded using the network channel). The host then sends out an address message on the broadcast channel and any node with the same key is set into receive mode; those whose key is invalid are reset. The host then sends any number of data messages, each of which will cause the local microprocessor to be interrupted and the data to be read. This process continues until such time as the local node receives a non-valid address. The system is particularly suited to downloading large quantities of data in parallel, such as software, to the remote nodes.

In addition to the direct addressing scheme for the broadcast channel, there is a masked addressing scheme in which only part of the address is compared with the key. This permits nodes to be grouped by their keys so that they may be either addressed singly or as one of a small number of groups. This mode is particularly useful for addressing layers in parallel or for accessing groups of input or output nodes, A natural result of this scheme is that a fully masked address is interpreted as valid by all nodes, regardless of their keys, and so may be used for executive functions, including bootstrap and a soft reset.

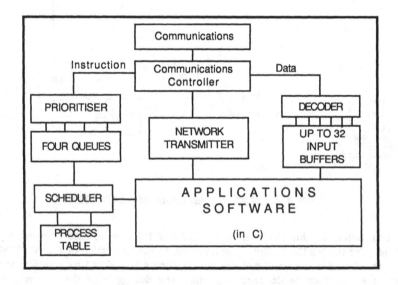

Figure 4 The major functional blocks of the local operating system

THE SOFTWARE ENVIRONMENT

The software environment splits into two sections: the host interface and the local operating system (figure 4). The latter is the more complex as it has to be efficient if the microprocessor is not to limit the simulation phase, yet still offer the flexibility to incorporate other user-defined routines which are not time-critical with a minimum of programming effort (such as diagnostic and set-up routines).

The front-end of the operating system is the communications integrated circuit since this is where all instructions will appear. The instruction handler receives messages over either the broadcast or the network channel and decodes them into instructions or data. The former are then added to instruction queues according to the priority of the instruction while the latter are added to data buffers which are associated with the relevant applications programs. The priority scheme permits the system to cope with background tasks as well as to interrupt operations for diagnostic purposes etc.

The main part of the memory in the local node is divided into a series of subroutines, each of which has its own parameter and data areas. The machine functions by receiving instructions from the host which are decoded into calls to these functions. A number of functions are provided (such as downloading and housekeeping routines) as well as the main simulation routines. This organisation permits different languages to be used to code the routines, so that time-critical functions may be optimised for speed whereas higher level languages may be used for more complex and less time-consuming functions such as analysis of the learning procedure. Further routines may be added at will until the local memory is exhausted.

In general, diagnostic routines will be resident in the remote nodes and only the results of the computation transferred to the host. Therefore, a simple method of developing code on the host and then downloading it for implementation in the remote nodes is essential.

The specification of the host software requires it to be the global coordinator of the system as well as being the interface to the user. Application programs are typically written in C using the normal facilities offered by the language and the specialist run-time library supplied. In this respect, the GRIFFIN may be thought of as a co-processor to the host.

NETSIM applications are developed on the host using a PC-based version of the NETSIM. When complete, they may be downloaded to the nodes. Corresponding software may be developed for the host to provide a customised interface to these routines. In a neural network simulation, this interface will normally consist of such functions as the definition of the interconnection map, the sigmoid function to be used and the runtime parameters.

The entire package is organised in a multitasking windows environment, permitting the user to monitor a number of processes concurrently; for example an error log, the progress of the simulation and intermediate diagnostic results.

CONCLUSIONS

The GRIFFIN offers a low cost environment in which to simulate large systems of neural networks at high speed. The increase in performance over a number of other parallel processing machines has been achieved by implementing specific functions in silicon while maintaining the flexibility of an industry-standard microprocessor. The operating system allows user applications to be developed in the host using high level languages and then downloaded directly to be run in parallel in the GRIFFIN. The machine offers performance equivalent to that of a supercomputer for neural networks applications but at a greatly reduced cost.

REFERENCES

Garth, S. C. J. "A chipset for high speed simulation of neural network systems", in *Proc IEEE 1st Intl. Conf. on Neural Networks*, San Diego, vol 3, pp.443–452, 1987.

Hillis, D., *The Connection Machine*. MIT Press, 1986.

Jackal, L. D., Howard, R. D., Graf, H. P., Straughn, B., Denker, J. S., "Artificial neural networks for computing" in *J. Vac. Sci. Technol.*, vol B4, p61, 1986.

Index